高等学校人文素养教育系列教材

生态文明与环境历史英语教程

梁艳春　祖国霞　主编

中国林業出版社
CFPH China Forestry Publishing House

图书在版编目(CIP)数据

生态文明与环境历史英语教程 / 梁艳春, 祖国霞主编. -- 北京 : 中国林业出版社, 2025. 1. -- (高等学校人文素养教育系列教材). -- ISBN 978-7-5219-3129-7

Ⅰ. X321.1

中国国家版本馆 CIP 数据核字第 2025V7V005 号

策划、责任编辑:范立鹏
责任校对:苏 梅
封面设计:周周设计局

出版发行 中国林业出版社(100009,北京市西城区刘海胡同 7 号,电话 010-83143626)
网 址 https://www.cfph.net
印 刷 北京中科印刷有限公司
版 次 2025 年 1 月第 1 版
印 次 2025 年 1 月第 1 次印刷
开 本 787mm×1092mm 1/16
印 张 13.25
字 数 320 千字
定 价 48.00 元

《生态文明与环境历史英语教程》编写人员

主　　编：梁艳春　祖国霞

副主编：欧　梅　杨惠媛

编写人员：（按姓氏拼音排序）

郭　陶　梁艳春　龙　莺　欧　梅

杨惠媛　祖国霞

前　言

2022年，教育部办公厅等四部门联合出台了《关于加快新农科建设推进高等农林教育创新发展的意见》。该意见明确将生态文明建设列为高等农林教育的重点领域，强调要加大相关专业人才的培养力度，将生态文明教育贯穿于整个育人过程。意见还要求深入推进课程教学改革，采用线上线下混合式教学模式，实施研讨式、探究式、参与式等多种教学方法，以促进学生自主学习，提升他们发现问题和解决问题的能力。此外，意见还强调要建立多元化的考核评价体系，将过程性考核与结果性考核有机结合，综合运用笔试、非标准化答案考试等多种形式，着力培养学生的创新意识和创新能力，加快新兴涉农专业教材和新形态教材的建设。当前，高等农林教育正朝着更加综合、创新和实践的方向发展，以更好地服务于国家的农业农村现代化和生态文明建设。

北京林业大学作为“双一流”农林类院校，肩负着培养具有“爱林、知林、为林”素养的生态化建设人才的重要使命。在新时代高等教育变革的背景下，英语作为一门国际通用语言，是获取全球生态信息和知识的关键工具。高校公共外语教学的目标是培养学生在专业领域内运用英语进行有效交流的能力，同时注重提升学生的文化素养和全球视野，帮助他们深入理解并积极参与国际生态文明建设，以响应国家对生态文明建设人才的战略需求。为契合这一教育背景和发展趋势，我们团队结合国家战略需求和近年来的教学实践，精心编写了本教材。

本教材的编写紧密结合农林学科的国家战略需求与高等教育的教学目标，突出学术性、实践性和国际化的综合特色。教材内容围绕生态文明和环境历史议题，涵盖农业发展、森林变迁和生态系统演变等模块，旨在帮助学习者掌握专业知识，拓宽国际视野，提升语言实践能力。同时，教材还致力于帮助学生树立正确的生态观，全面认识人类活动对自然环境的影响，为建设可持续发展的社会贡献力量，培养能够推动实现“美丽中国”所需的高质量人才。

本教材以知识和语言生成任务为目标，将课文作为知识和语言材料的输入载体，提供相关主题的专业知识、核心词汇、功能表达以及篇章组织结构，帮助学生为完成产出任务做好准备。通过形式多样、行之有效的驱动性问题和练习设计，教材引导学生利用思维结构导图和所习得的英语词汇，围绕当今环境热点展开讨论，训练学生批判性地分析文本传递的价值和背后所蕴含的深层意义，最终实现“见、学、用”的无缝对接，全面提升学生的思辨能力、跨文化能力和国际传播能力，使他们成为具有家国情怀、全球视野和专业本领的生态文明建设者和接班人。

本教材由八个单元组成，每单元围绕一个明确的主题展开，内容涵盖了“环境历史导论、史前人类与环境、古代农业文明与环境、近代早期生物圈的变化、现代工业革命与环境、二十世纪以来的环境问题、当代环境运动与生态思想、迈向可持续发展的未来”等重要的生态环境问题。每单元由导语、全球视角和中国视角三个部分组成，旨在从多维度探讨主题，从而拓宽学生的国际视野，促进文明互鉴。各单元内容设计相互关联，相辅相成，注重引导学生从批判性思维和跨文化交际的角度对每个主题进行深入分析和思考。每单元包括以下栏目：

导语(Preparation)板块通过内容概览和教学目标，为学生提供预习指南，明确学习重点和预期成就。内容概览以简洁的语言介绍单元的核心内容，帮助学生梳理课文之间的内在联系，为深入学习奠定基础。教学目标详细列出单元的学习目标，涵盖知识掌握、技能培养和情感态度3个层面，使学生明确学习路径和预期成就。

全球视角(Global Perspectives)板块通过两篇主题课文，广泛而深入地介绍不同国家在特定历史时期面临的环境问题，内容设计注重启发性和实用性。主题激活，结合单元主题，利用辅助视频、音频或图片等多种媒介创设学习情境，激发学生兴趣和思考，为主课文的学习进行热身。课文理解，提供两篇主课文，通过具体的环境案例展示不同历史时期和地区的环境问题。课文内容丰富，兼具学术性和价值引领性，突破传统教材单一文化视角的局限。课后练习，促进学生将所学知识与语言技能付诸实践。练习分为3个部分：知识应用，设置一到两个难度递增的练习，引导学生梳理课文结构并深化理解；语言技能，通过多样化的练习形式，如看图猜词、简答题、判断正误和翻译等，强化学生的词汇、阅读和翻译能力；综合交际，围绕开放性问题，鼓励学生基于前两部分的学习内容，通过小组合作和讨论，分享对相关话题的见解。

中国视角(China's Environmental Story)板块重点介绍中国在不同历史时期的环境应对措施及其在环境保护领域的努力和贡献。结合全球视角，进一步拓展学生的知识和思考深度。包括以下小节：观点讨论，理解中国引导学生在阅读课文的基础上，围绕一系列价值导向性问题展开深入讨论，加深对中国环境故事的理解。融合输出，讲述中国板块利用阅读材料创造真实的交际环境，帮助学生理解并融入习近平生态文明思想，通过综合运用所学知识，表达对中国环境故事的见解。

本教材打破了传统上将技能训练和知识传授分开的做法，将语言学习与知识探索融为一体，展现出以下独特的风格：一是绿色未来，中外共融。通过全球视角板块，培养学生对人类命运共同体的深刻理解，激励他们积极参与全球生态合作，为建设美丽中国和推动全球绿色发展贡献力量。通过中国视角板块，引导学生在国际视野中深刻理解中国的环境发展路径和环境治理智慧，增强对中国智慧的自豪感和文化认同感，展现全球与中国视角的有机结合。二是价值引领，育人为本。本教材秉承立德树人的宗旨，立足于美丽中国建设的战略目标，致力于培养具有创新意识、全球视野和实践能力的复合型人才。通过内容设计和教学方法的有机结合，教材鼓励学生以生态文明为指导，积极探索可持续发展的路径，为推动社会绿色转型贡献力量。三是质量为重，锤炼精品。为确保教材的专业性、科学性、价值引领性，并在多元文化交融中促进中外生态文明的沟通与互鉴、传播中国生态文化，编写团队进行了广泛调研和深入阅读，定期召开专题会议，集中讨论教学实施中的问题并提出解决方案，最终编撰出具有创新特色的全新英文教程。四是知识输入，任务驱动。教材采用“输入为基础，输出为导向”的编写理念，以立德树人为根本任务，通过学术文章阅读、案例分析和情境讨论等方式，提升学生的专业知识储备和语言应用能力。同时，注重培养学生的批判性思维，使其能够在全球视野下对生态问题进行深入探讨。五是产出导向，实践为本。教材融入探究式、参与式和项目式学习方法，通过设计基于实际问题的学习任务，如生态系统治理模拟、可持续农业规划和环境保护辩论等，提升学生的实践能力和创新意识，助力其在未来职业中有效应对复杂生态问题。

在此，诚挚感谢北京林业大学外语学院、教务处和中国林业出版社的大力支持，特别感谢范立鹏编辑给予的悉心指导与帮助！由于编者水平有限，难免会出现一些疏漏和不足。我们恳请使用本教材的同行和读者提出宝贵意见，以便我们不断完善和提高教材的学术质量和使用价值。

编　者

2025 年 1 月 6 日

CONTENTS

UNIT 1

An Introduction to Environment and History

Part One Preparation

Unit Preview

Environmental history in the West emerged during the 1960s and 1970s, spurred by growing awareness of global environmental issues such as pesticide pollution, ozone depletion, and human-induced climate change. In response to these pressing concerns, historians sought to trace the roots of these modern problems by integrating knowledge from various scientific fields that had developed over the previous century. This interdisciplinary approach laid the foundation for a comprehensive understanding of the complex interactions between humans and their environments.

While environmental history became a formal academic discipline in the mid-20th century, the study of human interactions with nature has much deeper historical roots. Thinkers such as Herodotus, Plato, and Mencius had already observed the impacts of human activities on nature and offered relevant insights. During the Middle Ages, local historical records documented environmental changes. In the modern era, George Perkins Marsh was one of the earliest forerunners to warn about environmental degradation caused by human actions, further solidifying the intellectual foundation of environmental history.

Just as Western scholars have long recognized environmental changes, China has an equally rich tradition of environmental documentation spanning centuries. Historical records from the Han, Tang, and Qing Dynasties provide valuable insights into environmental transformations. Given China's rich historical records and diverse ecosystems, there is significant potential for further development in this area.

Learning Objectives

Upon completion of this unit, you will be able to:

- understand the core themes and methodology of environmental history.
- understand the historical evolution of environmental thought and practice.
- explore how the Qing Dynasty documented and responded to environmental changes.

Part Two Global Perspectives

Active Reading 1

Warming Up

Task ***Discuss the following questions with you partner.***

1) Have you ever come across the English idiom "as dead as a dodo"? If so, how would you interpret its meaning?
2) What is the story behind it?
3) What lessons can we draw form the story?

Reading

What is Environmental History?

1 Environmental history is both one of the oldest and newest fields within human history. It is history focusing on human interaction with the natural world or, to put it in another way, it studies the interaction between culture and nature. A core topic in environmental history is how different peoples at different times have used, perceived, managed, and conserved their environments. The human species is part of nature, but compared to most other species we have caused far-reaching alterations of the conditions of land, sea, air, and the other forms of life that share our tenure of the Earth. The changes humans have made in the environment have in turn affected our societies and our histories. Environmental historians tend to think that the unavoidable fact that human societies and individuals are interrelated with the environment in mutual change deserves constant recognition in the writing of history. The environmental problems that received world attention during the last 40 years of the twentieth century, and whose importance has only increased in the present century, show the need for environmental histories that will help in understanding ways that humans have in part caused them, reacted to them, and attempted to deal with them.

2 Donald Worster, one of the leading figures in the field of environmental history, has contributed much to its development and methodology, and has recognised three clusters of issues to be addressed by environmental historians. The first cluster deals with the human intellectual realm consisting of perceptions, ethics, laws, myth and the other mental constructions related to the natural world. Ideas about the world around us influence the way we deal with the natural environment. Here we enter the second level of issues to be

studied: the level of the socio-economic realm. Ideas have an impact on politics, policies and the economy through which ideas materialise in the natural world.

3 But the world is not static, so it reacts to our actions to influence the material world. With the impact of human actions, we enter the third level of environmental history regarding the natural world. This level deals with understanding nature itself, the natural realm. In the case of woodland history, it is the way forest ecosystems have been working in the past and how they were changed by human actions. The impact of human actions on the natural world is causing a feedback that changes our ideas, policies, economy etc. In this way the natural world defines the limits of what we can do, and what not. Within this framework we try to change reactions we do not like and continue practices which, in our view, are successful. This model of the interaction between man and the environment depicts the concept of the separation between humans and nature. Although this division between the human and the natural realms is an artificial one, it can be a useful tool for the environmental historian in identifying important questions, the sources that might be able to answer the questions and the methods used to study these sources.

4 Environmental historians approach their field from a variety of perspectives because of its vast scope and complexity. Here we take American environmental history as an example to illustrate how historians conduct their study. According to Carolyn Merchant, a distinguished historian of environmental history, one approach is to focus on biological interactions between humans and the natural world. Animals, plants, pathogens, and people form an ecological complex in any one place that can be sustained or disrupted. When Europeans settled in North America and other temperate regions of the world, they introduced diseases, such as smallpox, measles, and bubonic plague; livestock, such as horses, cattle, and sheep; European grains such as wheat, rye, barley, and oats, along with varmints, such as rats; and weeds, such as plantain and dandelions. These ecological introductions, especially diseases, devastated the lives of native peoples. While some of the introductions, such as the horse, gave some Indians a temporary advantage over Europeans, the introduced ecological complex as a whole altered the landscape in ways that benefited the settlers and disrupted Indian lifeways.

5 A second way to think about environmental history is in terms of a series of levels of human interactions with nature, such as ecology, production, reproduction, and ideas. On the first level is nature itself. Nature's own history can be described in terms of the evolution of the geology and biology of a given place; the ecological succession of plants and animals found there; and the variations in temperature and climate that create the potential for human systems of production. At the second level, human forms of production also vary over time. North American Indians evolved complex systems of gathering, hunting, fishing, and horticulture, combined with trading across tribal boundaries. European settlers who arrived on the continent in the sixteenth and seventeenth

centuries developed sophisticated technologies such as ships, gunpowder, iron tools, clothing, and agricultural systems that created complicated, often uneven systems of interaction among Europeans, Indians, and nature. At the third level is reproduction. This includes biological and social forms of reproducing human and non-human life, as well as means of reproducing human social and political life over time. Finally, on the fourth level, are ideas, such as narrative, science, religion, and ethics that explain nature, the human place within it, and means of behaving in relation to it.

6 A third approach to doing environmental history is in terms of environmental politics and transformations in political and economic power. The history of the conservation and preservation movements of the late nineteenth and early twentieth centuries, for example, can be delineated in terms of political struggles within a presidential administration, the role of citizen movements in pressing for the preservation of natural areas, and the creation of government and state agencies to manage and conserve natural resources.

7 A fourth approach to the field is to focus on the history of ideas about nature. Histories of a philosophical idea such as wilderness, a scientific idea such as ecology, or an aesthetic idea such as natural beauty form the topics of numerous books about the nature of nature in North America. These works examine the ideas and creative products of artists, nature writers, science writers, explorers, and travelers for clues as to how people felt about nature and how their feelings led to actions with respect to its visual or economic resources. Such intellectual histories help us to understand how changing ideas about nature and beauty can be influential in creating the environments we see around us today.

8 A fifth way to do environmental history is in terms of narrative. One can argue that all peoples interpret their world through stories, whether the origin stories of Native Americans or Europeans, the stories of various fields of science as they progress over time, or morality stories that tell us how to behave in the world. Environmental historians often contrast their histories with histories of progress and enlightenment, inasmuch as developments for human well-being—such as industrialization—often result in the degradation of the environment through pollution and depletion. Yet all environmental history does not necessarily view history as a decline from a pristine environment that was irrevocably and negatively transformed when humans entered it. Environmental historians write narratives that are both progressive and declensionist, comic and tragic, intricate and bold. Nevertheless, the stories have a message. They explain the consequences of various past interactions with the natural world and warn us of potential problems as we form policies and make decisions that affect our lives and those of our children. Knowing and doing environmental history is therefore critical to the continuance of life on earth, whether that life be human or that of the other animals and plants that occupy the landscapes in which we dwell.

(Adapted from Merchant, 2005)

New Words and Expressions

perceive /pəˈsiːv/ *vt.* (以某种方式)看待,理解
tenure /ˈtenjʊə/ *n.* 使用权
cluster /ˈklʌstə/ *n.* 团,组
realm /relm/ *n.* 领域
ethics /ˈeθɪks/ *n.* 道德规范
static /ˈstætɪk/ *adj.* 静止的,停滞的
scope /skəʊp/ *n.* 范围
pathogen /ˈpæθədʒɪn/ *n.* 病原体
temperate /ˈtempərət/ *adj.* 温带的
measles /ˈmiːzlz/ *n.* 麻疹
bubonic plague /bjuːˌbɒnɪk ˈpleɪg/ *adj.* 鼠疫
rye /raɪ/ *n.* 黑麦
barley /ˈbɑːlɪ/ *n.* 大麦
oat /əʊt/ *n.* 燕麦
varmint /ˈvɑːmɪnt/ *n.* 有害动物
plantain /ˈplæntɪn/ *n.* 车前草
dandelion /ˈdændəlaɪən/ *n.* 蒲公英
devastate /ˈdevəsteɪt/ *vt.* 严重破坏,彻底摧毁
horticulture /ˈhɔːtɪˌkʌltʃə/ *n.* 园艺学,小型种植
tribal /ˈtraɪbəl/ *adj.* 部落的
delineate /dɪˈlɪnɪeɪt/ *vt.* 说明,解释
wilderness /ˈwɪldənɪs/ *n.* 荒野
aesthetic /iːsˈθetɪk/ *adj.* 美学的
narrative /ˈnærətɪv/ *n.* 叙事
enlightenment /ɪnˈlaɪtnmənt/ *n.* 启蒙
pristine /ˈprɪstaɪn,pristiːn/ *adj.* 未受损害的,处于原始状态的
irrevocably /ɪˈrevəkəblɪ/ *adv.* (决定、行动)不可更改地,不可撤回地
declensionist /dɪˈklenʃənɪst/ *n.* 衰败论叙事体
dwell /dwel/ *vi.* 居住

Exercises

Section I　Knowledge Focus

Task 1 ***Read Paras. 1-3 and complete the following diagram by filling in the blanks with the letters corresponding to the words given below. You may use any letter more than once.***

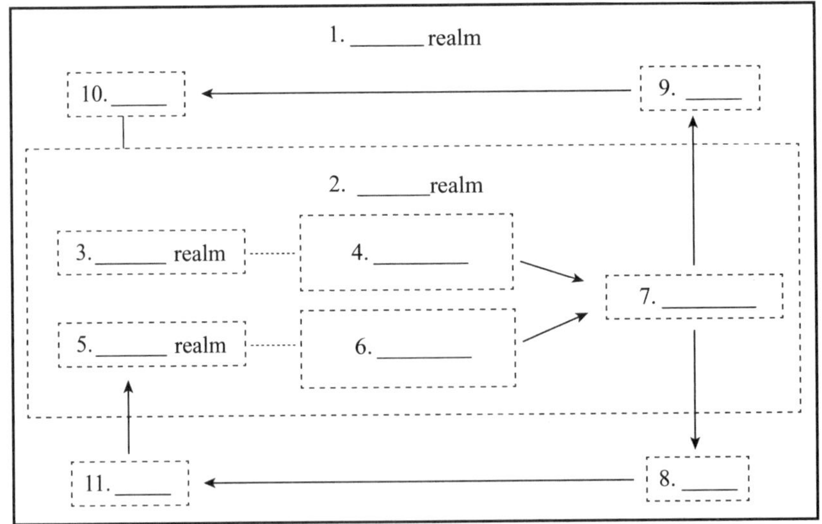

A. impacts
B. natural
C. perceptions, ethics, law, and myth
D. socio-economic
E. human actions
F. feedback
G. politics, policies and the economy
H. intellectual
I. human

Task 2 ***Read Paras. 4-5 and choose the correct heading for each statement from the list of headings given below.***

A. Production B. Ecology C. Biological interaction D. Ideas E. Reproduction

__________ 1) The biological and social processes through which humans are born, nurtured, socialized, and governed.

__________ 2) The way in which the world is perceived, understood, and interpreted at any given time by a society or social group. Individual consciousness is the totality of one's thoughts, feelings, and impressions—the awareness of one's acts and volitions.

__________ 3) It examines the biotic interchange ranging from living to non-living beings in the natural environment, which could be mutualism (beneficial to both partners) to competition (harmful to both partners).

__________ 4) The extraction, processing, and exchange of nature's parts as resources, whether as gifts in traditional economies, as bartered goods and services in subsistence economies, or as commodities in market economies.

__________ 5) It studies nature itself and its changes over time, as well as the effects of those changes on humans.

Task 3 ***Given below are five statements. Each statement contains information from one of the paragraphs of the text. Identify the paragraph from which the information is derived. Write down the paragraph number (1-8) for each statement.***

__________ A) Knowing about people's varying perceptions of natural environments can help us picture the world around us at present.

__________ B) As individuals and societies, we make choices about how we use resources and respond to new risks. These choices are made through the lens of economic value and political institutions.

__________ C) Environmental history sometimes can serve as a corrective to the prevalent tendency of humans to see themselves as separate from nature.

__________ D) Storytelling-based research on environmental history can have a positive impact on people's environmental awareness.

__________ E) Environmental history is a kind of history that seeks understanding of human beings as they have lived, worked, and thought in relationship to the rest of nature through the changes brought by time.

__________ F) Environmental history considers the environment itself and its effects on humans.

__________ G) Reactions to ecological introductions have been shown to have mixed results.

__________ H) Donald Worster has proposed three levels of analysis that environmental history should seek to cover.

Section II Language Focus

Task 4 ***Complete the crossword puzzle using the clues given. Most of the answers can be found in this text.***

1) A statue and a paused video have this in common.
2) What do wise people do with water during a drought?
3) How does a healthy diet affect your life?
4) A protected area for endangered animals.
5) This type of climate is neither too hot nor too cold.
6) The world where kings rule, or where elves and dragons roam.

Task 5 ***Complete the following sentences with appropriate words or expressions given below. Change the form where necessary.***

realm sustain temperate conserve perceive

1) He is interested in how our __________ of conservation movement affect the way we live.
2) In the conclusion of his speech, it appeared as though he was moving into the __________ of fantasy.
3) The local community has initiated a series of __________ practices in order to promote ecological balance within the region.
4) According to one estimate, tropical rainforests contain around half of all the world's species far more than the __________ forests of Europe and America.
5) On a __________ estimate, there are now about 5,000 books or articles that deal with the topic of climate change, at least in part.

artificial ethics constant devastate tribal

6) He's always had good work __________, which is why we're starting to see good results from him.
7) When they had a battle, the leader of the losing __________ would be killed and eaten by the winners.
8) A boycott would __________ a big chunk of the country's farming and cause job losses.
9) Some islands are __________, such as the obscenely expensive projects of Qatar's Pearl resorts or Dubai's Palm Islands.
10) In our __________ changing, global, highly technological society, collaboration is a necessity.

Task 6 ***Match the words in the left column with those in the right column to form appropriate expressions. Then complete the following sentences with one of the expressions. Change the form or add articles where necessary.***

pristine	rites
temperate	force
aesthetic	zone
tribal	wilderness
static	appeal

1) Artists traveled up the Hudson in search of __________, documenting the remnants of a natural world that was fast disappearing.
2) Reptiles of the north __________ include many ecological types.

3) Early conservation viewed nature as a ___________, but modern ecology recognizes ecosystems as dynamic and constantly evolving.

4) The museum, it said, lacked ___________, organizational coherence, and the perception of substantive balance.

5) In some Latin-American countries, marriages performed under indigenous ___________ are not recorded as legal.

Task 7 ***Translate the following Chinese expressions into English with what you have learned from the text.***

远在欧洲殖民者抵达美洲之前，印第安人以狩猎、采集、种植和渔业为生，在北美洲多种多样的环境中发展出了各种复杂的文化、社会组织和语言系统。然而，欧洲殖民者的到来改变了当地的文化和生态系统，印第安人的土地被夺走，文化被侵蚀，人口因疾病和战争而急剧减少。这个过程中，许多印第安人群体遭受了极大的苦难。尽管如此，印第安人的文化和传统在一些社区中得以保留和传承。

Section III Sharing Your Ideas

Task 8 ***Environmental history studies the long-term interaction between humans and the natural world. Could you give some specific examples of the interplay between human activities and the environment? Share your examples with the class.***

Active Reading 2

Warming Up

Task ***Discuss the following questions with your partners.***

1) Can you name some major environmental issues we face today?

2) How do you think these issues were viewed or handled in the past?

3) Why do people study environmental history?

Reading

Forerunners of Environmental History from Ancient to Modern Times

1 Environmental history as a conscious exploration of human relationships to the natural environment in the past, that is to say, as a historical discipline, began in the late twentieth century and is one of the newest scholarly endeavors. But the questions asked by environmental historians are in many cases old ones that attracted the interest of writers from the Greeks and Chinese among other ancient peoples, through the centuries down to modern times.

2 The first Greek historian whose work survives, Herodotus, recorded a number of

remarkable changes made in the natural environment by human efforts, and generally reported negative consequences from them. He believed that massive works like bridges and canals demonstrated an overreaching human pride that might call forth punishment from the gods. He wrote that when the Cnidians started to dig a canal through the neck of land that connected their city to the mainland in order to improve their defenses, the workmen suffered an unusual number of injuries from flying splinters of rock. Wondering why this was happening, they sent an embassy to ask the oracle at Delphi, who replied in direct words rather than her customary riddle: "Do not fence off the isthmus; do not dig. Zeus would have made an island, had he willed it." The command to put down tools and cease work was duly issued.

3 Plato included advice concerning environmental problems for his ideal states in the *Republic* and the *Laws*. He also observed historical deforestation of the mountains of Attica in the *Critias*, offering archaeological evidence: large roof beams in huge buildings that still stood in his own day had been cut from mountains where only "food for bees" (flowering herbs and bushes) remained. The former forest had served to store and release the rains, producing many springs, as evidenced by shrines that stood at places where springs had been, though they were dry in Plato's day. In the same period as deforestation, massive erosion had removed the rich, soft soil, leaving only the rocky framework of the land, which Plato compares to the skeletal body of a man wasted away by disease.

4 An apt comparison may be made between Plato and the Chinese philosopher Mencius, who also lived in the fourth century BCE and described deforestation in his homeland. Mencius saw a mountain that had been denuded of its forests over the years by logging, and the way in which grazing can make deforestation permanent by preventing reproduction and the growth of small trees. As a follower of Confucius, Mencius made many interesting comments on the human relationship to nature and gave some valuable advice on land management, which is an important topic that he considered to be one of the primary responsibilities of the state. He insisted that "the people are of supreme importance; the altars to the gods of earth and grain come next; last comes the ruler". In theory, the ruler owned the land and parceled it out to those who used it, but rulers were not exempt from labor on behalf of the altars or the people. A landlord had to plow the land to grow grain for sacrifices. The condition of the environment in a country offered telling evidence concerning the merit of its government.

5 To the environmental historian, a distinctive emphasis of Mencius is his recommendation of conservation practices to assure that resources would not be exhausted, but could feed the people from year to year. He grasped the principle of sustainable use of renewable resources. Speaking to King Hui, he once advised, the people should be allowed to work in the fields at seed time and harvest, not marched off to war. Nets with wide mesh to be

used in fishery would allow small fish and turtles to escape and grow to catchable size. A form of sustained-yield forestry would assure a supply of wood in succeeding years. Mencius' advice concerning forest conservation was particularly sound.

6 Information about environmental change in the Middle Ages is more likely to come from local histories than from general histories, since such changes were more often noted in the landscape of a single district. In *Green Imperialism*, one of the groundbreaking studies of Richard Grove, which has fused colonial expansion, the roots of modern science, and what Richard calls the roots of environmentalism in a multi-layered attempt, shows that scientists, including physicians, sent out by colonial powers as early as the seventeenth century, noticed environmental changes on oceanic islands and in India and South Africa, changes so rapid that they could be chronicled within the span of a human life. They recorded evidence of human-induced deforestation and climate change.

7 Among modern authors who helped turn attention to environmental history is George Perkins Marsh, who long served as US ambassador to Italy. He observed in the Mediterranean area and elsewhere "the character and extent of the changes produced by human action in the physical condition of the globe we inhabit", and warned in his great work, *Man and Nature*, published in 1864, that "the result of man's ignorant disregard of the laws of nature was deterioration of the land". *Man and Nature* was intended to be a worldwide survey of the ways in which humankind had damaged nature and continued to do so; for him, Rome was not the only organized society that experienced environmental crisis, but his familiarity with the Mediterranean countries, Europe, and North America led to an emphasis on those areas. Marsh may be regarded as the first of the precursors to environmental history who systematically investigated the question of environmental deterioration and the possible exhaustion of natural resources.

8 The early twentieth century saw the rise of interest in what was then called conservation history, which was concerned with the Progressive Conservation Movement and questions such as land use, resource conservation, and wilderness. The emergence of environmentalism after mid-century meant that historians turned their attention also to issues such as pollution, lifestyles, and environmental legislation.

9 Environmental history first emerged as a conscious historical effort in the United States in the 1960s and 1970s, where it was named and organized as a distinct subdiscipline of history (this statement is not intended to deny that many of the themes of environmental history had already emerged in the works of European historians). At that time, historians had already noted the conservation movement in America, which included advocates of nature preservation, amongst them the Progressive Conservation Movement and John Muir who founded Sierra Club (an American environmental organization), which campaigned on behalf of prudent and scientifically based use of natural resources. The Progressive

Conservationists received the powerful support of the White House during the administrations of Theodore Roosevelt (1901–1909) and Franklin D. Roosevelt (1933–1945).

10 It was Samuel Hays, who defined the great change in American attitudes toward the environment in the period that gave birth to the environmental movement and to environmental history as a scholarly endeavor. In an article, "From Conservation to Environment: Environmental Politics in the United States since World War Ⅱ", later expanded in a book, *Beauty, Health, and Permanence*, Hays noted the emergence of new environmental values, including the desire for environmental amenities, recreation, aesthetics, and health, all associated with rising standards of living and education. Of course, Americans had been camping and hiking and enjoying the out-of-doors for at least half a century. In the 1950s, free from preoccupations of economic depression and war, they sought environmentally related recreation in unprecedented numbers.

11 Americans were also increasingly concerned with environmental issues that directly affected them, beyond land use and resources. They became aware of the dangers of radioactive contamination by fallout from nuclear bomb tests; news media told them of oil spills and water pollution in the Great Lakes; there were gasoline shortages across the country; and from the cities to the Grand Canyon people could see and feel higher levels of air pollution. In her 1962 book *Silent Spring*, Rachel Carson warned of damage from persistent pesticides, and emerging environmental movements reached nationwide awareness on the first Earth Day, April 22, 1970. Following this, a series of environmental laws was enacted by Congress and signed by several presidents, including Richard M. Nixon. Ecology, formerly a little-known science, became a household word.

12 It is undoubtedly the case that the historians who created the field of environmental history in the 1960s and 1970s were for the most part environmentalists. In 1976, American Society for Environmental History (ASEH) was formed by a group of scholars, mostly historians and environmentalists. The journal of the society began publication in the same year, and was successively titled *Environmental Review* (1976–1989), *Environmental History Review* (1990–1995), and *Environmental History* (1996–present). Nonetheless, environmental history has remained from its conception an interdisciplinary endeavor everywhere that it is practiced.

(Adapted from Hughes, 2015)

New Words and Expressions

endeavor /ɪnˈdevə/ *n.* 努力，尽力
overreaching /ˌəuvəˈriːtʃɪŋ/ *adj.* 不自量力的
splinter /ˈsplɪntə/ *n.* 碎片，裂片，尖片
oracle /ˈɔrəkəl/ *n.* 神谕，神示
customary /ˈkʌstəmərɪ/ *adj.* 习惯的
isthmus /ˈɪsməs/ *n.* 地峡
cease /siːs/ *vt.* 终止，停止
duly /ˈdjuːlɪ/ *adj.* 适当的，适时的
archaeological /ˌɑːkɪəˈlɔdʒɪk(ə)l/ *adj.* 考古学的
shrine /ʃraɪn/ *n.* 圣地，神殿
skeletal /ˈskelətəl/ *adj.* 骨瘦如柴的
apt /æpt/ *adj.* 恰当的
denude /dɪˈnjuːd/ *vt.* 伐光（森林），剥夺
supreme /sʊˈpriːm/ *adj.* 至高无上的
altar /ˈɔːltə/ *n.* 神坛，祭祀
mesh /meʃ/ *n.* 网，网状物
succeeding /səkˈsiːdɪŋ/ *adj.* 随后的，接着的
groundbreaking /ˈgraʊndˌbreɪkɪŋ/ *adj.* 开创性的
fuse /fjuːz/ *vt.* 融合，结合
chronicle /ˈkrɔnɪkəl/ *vt.* 按时间顺序记载
induce /ɪnˈdjuːs/ *vt.* 引诱，诱导
disregard /ˌdɪsrɪˈgɑːd/ *vt.* 忽略，弃置
deterioration /dɪˌtɪərɪəˈreɪʃən/ *n.* 恶化
precursor /prɪˈkɜːsə/ *n.* 先驱，先兆
prudent /ˈpruːdənt/ *adj.* 谨慎的，慎重的
preoccupation /prɪˌɔkjuˈpeɪʃn/ *n.* 使人全神贯注的事物
unprecedented /ʌnˈpresɪdəntɪd/ *adj.* 前所未有的，史无前例的
fallout /ˈfɔːlaʊt/ *n.* （核爆炸后的）放射性坠尘
spill /spɪl/ *n.* 泄漏
persistent /pəˈsɪstənt/ *adj.* （化学品，放射性）作用持久的，挥发慢的
pesticide /ˈpestɪsaɪd/ *n.* 杀虫剂

Exercises

Section I　Knowledge Focus

Task 1 ***Based on your understanding of Paras. 1-3, decide whether the following statements are true or false. Put T for true and F for false in the blank provided before each statement.***

__________ 1) Plato's works *The Republic* and *The Laws* include no discussion on environmental issues.

__________ 2) Although environmental history is a relatively newly-recognized discipline in recent decades, it has long been a subject of interest among numerous scholars since ancient times.

__________ 3) Herodotus depicted that when people went to consult the gods, they received an oracle that was rather obscure and difficult to understand.

__________ 4) Plato was a keen observer of environmental changes happening around his surroundings.

__________ 5) Herodotus believed that damnation would be the consequence of disturbing the order of nature.

Task 2 ***Read Paras. 4-9 and answer the following questions by choosing the most appropriate answer.***

1) Which one below is not the reason why the author thinks Plato can be compared to Mencius?
   A. They were contemporaries.
   B. They were both great philosophers.
   C. They both had observations of environmental changes in their works.
   D. They both maintained a religious attitude towards the environment.

2) Which of the following statements about Mencius is true according to the text?
   A. He advocated that the rule is the most significant entity in the world.
   B. He believed that a governor who cares for the Earth and can cope with environmental problems can be trusted to govern well.
   C. He was advised by King Hui of Liang that the wise use of natural resources is essential for thriving and sustaining life for future generation.
   D. He suggested that the people should be working in the fields rather than going to the battlefield.

3) According to paragraph 6, what influence may Richard Grove have on environmental history?
   A. He was one of the founders of the environmental history.
   B. He made a major contribution to studying the roots of modern environmental concern.
   C. He had a deep passion for colonial history.
   D. He was the first to provide impetus for the idea that humans have caused environmental alterations around the world.

4) What can you infer from the text about George Perkins Marsh?
   A. He was the first-ever U. S. Minister to raise concerns about the environment.
   B. He specializes in Roman environmental history.
   C. He believed that environmental crisis was entirely the result of people's personality.
   D. He suggests that human beings were agents of change.

5) Which of the following statements is True?
   A. Conservation history gave birth to environmentalism.
   B. The focus of scholars has shifted over the course of the 20th century.
   C. A series of environmental laws were passed before the mid-1950s.
   D. European historians have no interest in environmental history.

Task 3 ***Read Paras. 8-12 and complete the following diagram by filling in the blanks with words from these paragraphs. Change the form where necessary. (ONE WORD ONLY)***

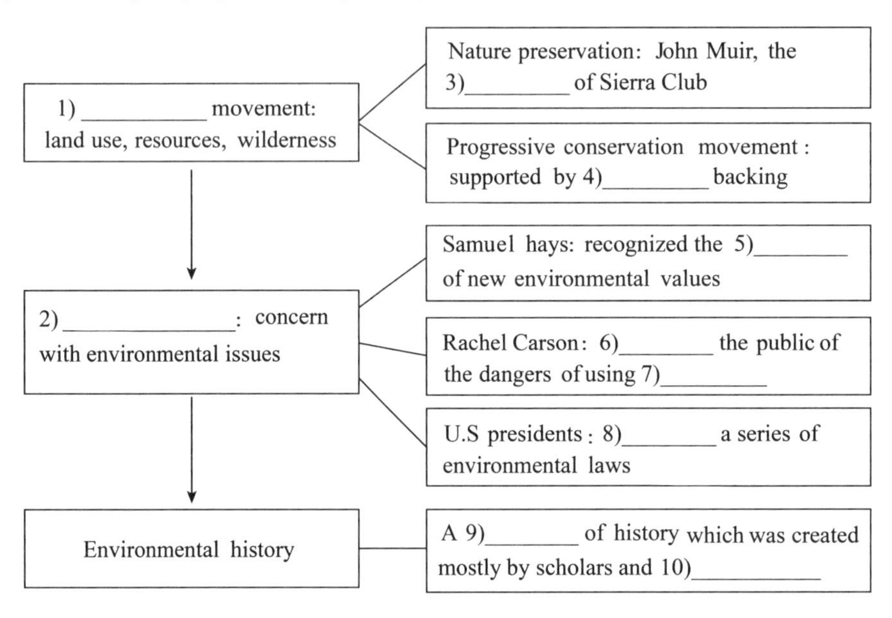

Section II Language Focus

Task 4 ***Watch a video clip about conservation movement and then fill in the blanks based on what you hear.***

The European explorers who came to North America in the early 16th century were driven by different motivations, while they all shared a desire for wealth. To do this, they established the profiteering farming, 1) __________, and mining economies, which eventually disrupted the local ecosystems. With the rapid 2) __________ of vast American 3)__________, it was not until the late 1800s when some Americans began to recognized that resources in the natural were not 4) __________ commodity that could be endless 5)__________. By the 1870s, advocates of 6) __________ started claiming that it is urgent to protect everything in the nature from wildlife to landscapes before it's too late.

It was the hard work of some devoted people that pushed forward the conservation movement. a few US presidents and John Muir were the most influential figures among them. John Muir, once was originally a master mechanic and inventor, but he 7)__________ on the technology after an industrial accident that caused him with eye injury. Then, for the rest of his life he 8)__________ time and effort to 9)__________ the beauty of nature and became one of the greatest 10)__________ in the history.

Task 5 ***Identify what is being described in the following phrases. Choose your answer from the box below and write the corresponding word in the space provided.***

apt customary sound cease merit

___________ 1) The great virtue of the project is its flexibility and low cost.

___________ 2) We hope that the military regime will desist from its acts of violence.

___________ 3) The timing of announcement was particularly appropriate.

___________ 4) Good clinical practice requires sensible decision-making and doing. the right test at the right time to guide patient care.

___________ 5) In some cultures, it is conventional for the bride to wear white.

Task 6 ***Complete the following sentences with the clues given in brackets. You may find them from the text and change the form where necessary.*** (***ONE WORD ONLY***)

1) Agricultural subsidies and a thoughtless ___________ (漠视) for natural processes are washing away the commonwealth of land, its soils and wildlife.
2) I believe we should support responsible organizations ___________ (宣传) on behalf of animals.
3) Federal and state engineers are seeking ways to capture and treat the ___________ (污染) runoff.
4) Successful cloning of the cactus could be used for revegetation of desert ___________ (砍伐) by fire or development.
5) Air quality is rapidly ___________ (破坏) in our cities.

Task 7 ***Translate the following expressions from Chinese into English with what you have learned from the text.***

1) a conscious exploration
2) scholarly endeavors
3) customary riddle
4) renewable resources
5) a multi-layered attempt
6) human-induced climate change
7) environmental amenities
8) unprecedented numbers
9) persistent pesticides
10) nationwide awareness

Section III Sharing Your Ideas

Task 8 ***Watch a video about Earth Day, and discuss the following questions with your partners.***

视听资源

1) What is the significance of Earth Day, and how is it observed?
2) What are the historical and contemporary reasons for celebrating Earth Day?
3) What daily actions can individuals undertake to promote environmental sustainability?

Part Three China's Environmental Story

Active Reading 3

Warming Up

Task ***Over the past decade, China has undergone tremendous transformations in various aspects, particularly in its natural environment. Describe to your partner the ways in which your hometown has changed over the last 10 years.***

Reading

Chinese Environmental History in the Qing Dynasty

1 Some awareness of what we would now call "environmental history" existed in the Qing Dynasty, which made policies aimed at protecting the habitats of valued vegetable and animal resources. For fifty years the Qing attempted to "rest the hills by rotating collection" in order to safeguard stocks of wild ginseng in northeastern China. The authorities likewise realized the effects of habitat destruction: "when the wastelands are gradually opened to farming, the prolongation of the existence of the sables and (other) animals is cut off."

2 Some Chinese of this period had a premonition of the coming exhaustion of resources, of the possible "wearing out of the world". An illustration of this is the poem by Wang Taiyue, *Lament for the Copper-bearing Hills* written in the mid-eighteenth century. As the ore grew scarcer and the wood needed for smelting was used up, winning a living became steadily harder for the miners:

> *They gather, at dawn, by the mouth of the shaft,*
> *Standing there naked, their garments stripped off,*
> *Lamps strapped to their heads in carrying-baskets,*
> *To probe in the darkness the fathomless bottom.*
> ...
> *The wood they must have is no longer available.*
> *The woods are shaved bald, like a convict's head. Blighted.*
> *Only now they regret-felling day after day.*
> *Has left them no way to provide for their firewood.*

3 Observers in late-imperial times also focused on the lack of environmental resilience of the

areas outside the traditional zone of Chinese economic exploitation when traditional farming technology was applied to them. Thus Fang Gongqian, who had been banished to Ninguta, wrote of the swift exhaustion of newly opened land in Jilin:

One can farm along the line of the hills... with no taxes levied. It costs dear[though] to develop this waste ground. When you hoe it the first year, it still remains waste. The second year it becomes mature, and in the third, fourth, and fifth years it is rich. In the sixth or the seventh year you abandon it, and hoe some other area.

4 The practical dilemmas that confronted the Qing authorities concerned with the people's livelihood are summed up in the *Song of the Timber Yards* by Yan Ruyi. He speaks both of the economic usefulness of logging and of the environmental destruction that it causes. He appreciates the employment that it affords, and fears the social disruption that can occur if it stops suddenly, as when wage costs, due to the relatively rising price of food grain, make it uneconomical to employ people.

5 From the side of historical geography, a number of the articles in works such as *Mirror Lake and water control in Shaoxing*, edited by Sheng Honglang, may be classified as environmental history, in this particular case covering anthropogenic changes in the hydrology of the region. As regards particular topics, Tan Qixian, who is best known as a historical cartographer, has, for example, opened up a debate concerning the effect of the historical removal of the vegetation cover in the middle reaches of the Yellow River on the density of the river's sediment loading, and hence on the variable historical frequency of the breaching of dikes in the lower reaches caused by increased deposition on the bed of the river. As regards the pioneering of regionally focused studies, Chen Qiaoyi has likewise laid the foundations of the environmental history of northern Zhejiang, documenting the stages of the loss of forest cover and changes in crop use. His work also relates the rise and fall of particular hydraulic systems to their internal logic (notably their interactions with other nearby systems), to sedimentation due to both natural and man-made causes, and to the shifting balance of power between farmers concerned with water supply and those eager to reclaim lake-bottom lands.

6 A different sort of literature is exemplified by Liang Biqi and Ye Jinzhao's *Natural disasters in Guangdong*, which offers something of a contrast. The term "natural" in the title is perhaps a misnomer, since the authors also deal with anthropogenic damage to the environment, and with degradation and pollution that occur slowly, not just "disasters" as commonly understood. It is scientifically informed, at a level appropriate for general policy formulation, and is motivated by the concern that environmental damage, both natural and man-made, obstructs and holds back economic growth, especially if this growth is ill-conceived. Socio-natural causal feedback loops are identified and analyzed, an example being the spread of rats when cropping patterns are shifted toward fruit and

sugarcane (which they particularly favor), the natural enemies of rats are exterminated or made extinct, and the rats themselves evolve resistance to rat poisons. The historical section consists mostly of catalogs of events, but the crude data indicating a large long-term increase in the frequency both of droughts and of floods in Guangdong from Song to early modern times are prima facie evidence for continuing adverse effects from human interference with ecosystems. The true statistical significance of the trends in these data-in the light of a probably improved social recording system and a presumably increased exposure of the population as it spread into relatively less favorable habitats—is not, however, given the analysis it requires.

7 As environmental history develops as a newer discipline in China, these extensive studies are bound to hold a prominent place in it. Forming our own history of environment is not simply a far-fetched dream, but a realistic possibility that is supported by the numerous research foundations that have been established even further back in ancient times.

(Adapted from Elvin et al., 1998)

New Words and Expressions

prolongation /ˌprəʊlɒŋˈɡeɪʃən/ *n.* 延长，延续

sable /ˈseɪbl/ *n.* 紫貂

ginseng /ˈdʒɪnseŋ/ *n.* 人参

premonition /ˌpriːməˈnɪʃn/ *n.* (尤指不好的)预感

lament /ləˈment/ *n.* 挽歌，悼文；表达哀伤(或痛惜)之情的言辞

ore /ɔː(r)/ *n.* 矿石，矿砂

shaft /ʃɑːft/ *n.* 竖井，通风井

fathomless /ˈfæðəmləs/ *adj.* 深不可测的，不可了解的

convict /ˈkɒnvɪkt/ *n.* 罪犯，囚犯

blight /blaɪt/ *vt.* 破坏，使枯萎

banish /ˈbænɪʃ/ *vt.* 驱逐，发配(边疆)

levy /ˈlevɪ/ *vi.* 征收(罚款、税款等)

anthropogenic /ˌænθrəʊpəʊˈdʒenɪk/ *adj.* 人为的，由人类活动引起的

hydrology /haɪˈdrɒlədʒɪ/ *n.* 水文学

cartographer /kɑːˈtɒɡrəfə(r)/ *n.* 制图师

breach /briːtʃ/ *vt.* 违反，破坏

dike /daɪk/ *n.* 堤坝，大坝

deposition /ˌdepəˈzɪʃn/ *n.* 沉淀(物)

hydraulic /haɪˈdrɒlɪk/ *adj.* 与水利(学)相关的

sedimentation /sedɪmenˈteɪʃn/ *n.* 沉积(作用)

reclaim /rɪˈkleɪm/ *vt.* 开垦

exemplify /ɪɡˈzemplɪfaɪ/ *vt.* 是……的典范；举例说明

misnomer /mɪsˈnəʊmə/ *n.* 错误的名字(或名称)

exterminate /ɪkˈstɜːməneɪt/ *vt.* 根除，消灭

extinct /ɪkˈstɪŋkt/ *adj.* 灭绝的，消亡的

crude /kruːd/ *adj.* 粗略的，大概的

adverse /ˈædvɜːs/ *adj.* 有害的，不利的

interference /ˌɪntəˈfɪərəns/ *n.* 干扰，干预

Exercises

Section I Understanding the Text

Task 1 ***This reading passage has seven paragraphs. Choose the correct heading for each paragraph from the list of headings below.***

A. Agricultural challenges in underdeveloped regions
B. Environmental consciousness and conservation practices
C. A Paradoxical challenge faced by qing authorities
D. Insights from other disciplines contributed to environmental history in China
E. Prospects for environmental history in China
F. The correlation between natural catastrophes and human intervention
G. Concerns regarding the exhaustion of resources

Paragraph 1 ______ Paragraph 2 ______ Paragraph 3 ______ Paragraph 4 ______
Paragraph 5 ______ Paragraph 6 ______ Paragraph 7 ______

Section II Developing Critical Thinking

Mesopotamia, ancient Egypt, ancient India, and ancient China are among the earliest civilizations. Of these, China is the only one that has developed in an unbroken chain to the present day, whereas the others have collapsed for various reasons. China's unique and uninterrupted historical trajectory stands out globally.

Task 2 ***What factors contributed to China's resilience while other ancient civilizations declined? An analysis of the underlying reasons for China's historical endurance may include the following considerations:***

- Landscapes
- Agricultural practices
- Adaptation to environmental changes
- Cultural attitudes towards nature
- Technological innovations
- Long-term planning and governance

Mind-mapping and Sharing Session

Please form groups of two or three. Each group will be assigned a specific topic and will create a mind map (either on paper or digitally) to organize your ideas. Visual elements such as colors, symbols, and overall design are encouraged to enhance the map's appeal. Once your group has completed the mind map, you will present it to the class. Please follow these guidelines.

Step 1　Focus on your central topic and do extensive research

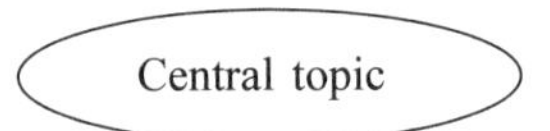

Step 2　Branch out the gathered information into subtopics

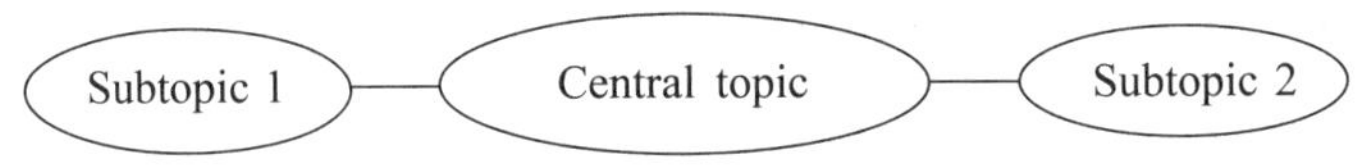

Step 3　Use keywords

Use keywords, phrases, or images instead of complete sentences to represent ideas on your mind map. This keeps the information concise and visually engaging(see Example 1).

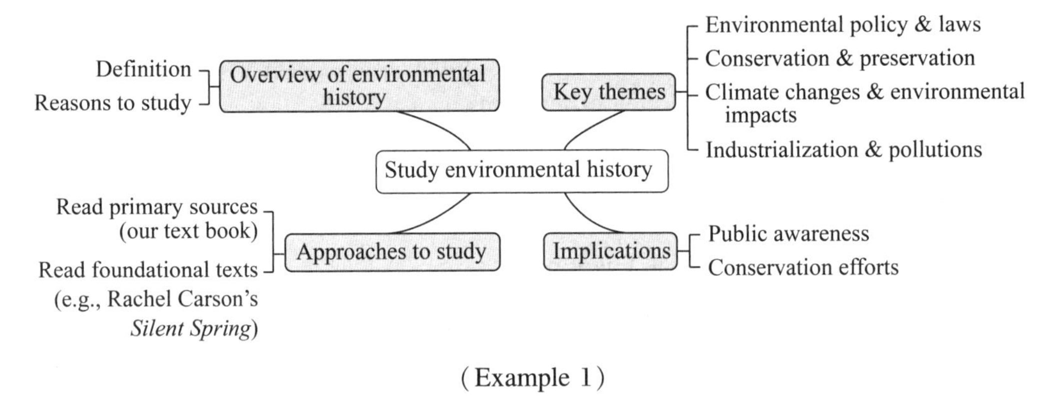

(Example 1)

Step 4　Identify connections between different ideas

Use lines or arrows to link related concepts(see Example 2).

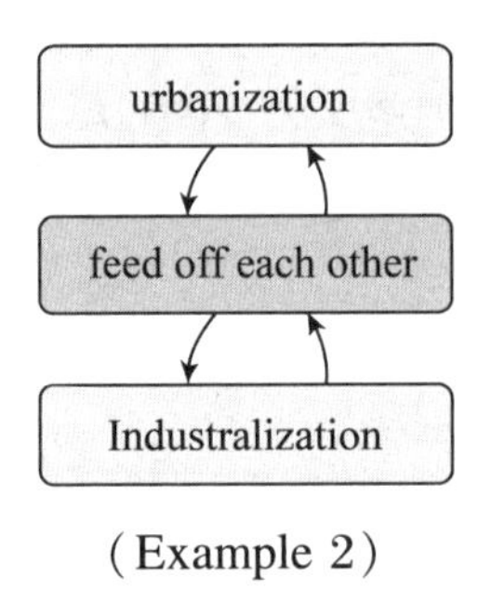

(Example 2)

Step 5　Enhance your mind map with color and visuals

Step 6　Self-check these questions before sharing to ensure a qualified mind map

1) Does the mind map cover the topic comprehensively, including relevant subtopics, details, and examples?
2) Are the information and data presented on the mind map accurate and factual?
3) Are the transitions between different sections or branches smooth and logical?
4) Does the mind map demonstrate creativity and originality in its design and presentation?

Step 7　Share your work

UNIT 2

Human-Environment Interactions in Early Times

Part One Preparation

Unit Preview

Throughout history, the way of human-environment interactions has undergone remarkable evolution. These interactions can be classified into three primary categories.

Dependence on the Environment: Humans have always relied on their environment for essential resources such as food, water, and shelter. Nowhere is this dependence more evident than in the case of Native Americans in North America. They carefully adjusted their activities in accordance with the seasons and the availability of resources, vividly demonstrating their deep-rooted dependence on the environment for survival.

Modification of the Environment: Humans frequently modify their environment to satisfy their requirements, which leads to a spectrum of impacts on the ecosystem, ranging from negligible to highly disruptive. The fate of Easter Island serves as a poignant reminder of the catastrophic consequences of unregulated environmental modification. It compels us to exercise greater caution in our actions to safeguard a sustainable future.

Adaptation to the Environment: Humans have adapted both genetically and culturally to thrive in a vast array of environments, spanning from the frigid Arctic Circle to the tropical equatorial regions. China stands as a prime example of human-environment adaptation, where the Chinese people, deeply influenced by traditional beliefs and practices, respect nature and apply traditional ecological wisdom for sustainable living.

These historical cases highlight how different societies adapted to and transformed their ecosystems, emphasizing the delicate equilibrium between human needs and environmental sustainability.

Learning Objectives

Upon completion of this unit, you will be able to:

- understand the seasonal and ecological adaptation strategies of Native American Indians.
- understand the environmental collapse of Easter Island and learn the lessons about resource management.
- evaluate various human-environment interactions in early times around the world.

Part Two Global Perspectives

Active Reading 1

Warming Up

Task ***Watch a short video clip. Then discuss the following questions with your partners.***

1) Why is the term "Indian" considered a mistake according to the video?
2) When and from where did the ancestors of the American Indians originally migrate, according to most of the archaeologists?
3) Why have most Native American Indian civilizations ceased to exist in the present day?
4) What else do you know about these indigenous people?

Reading

North American Indians and Nature

1 To survive in North America, the Indians exploited the seasonal diversity of the various landscapes they inhabited. This was especially true in the far northern reaches of New England, where cold and frost, in addition to the stony soil deposited by the glaciers, made agriculture a risky and difficult venture. Compelled to adopt hunting and gathering, Native Americans found that in a temperate climate, the spring and summer months offered a plentiful supply of food. From March through May, the Indians used nets, weirs, and canoes to catch fish on their way to spawn upstream, while migrating birds such as Canada geese and mourning doves further bolstered the food supply. In the summer months, they also gathered various kinds of nuts and berries. By the fall, however, as the temperature turned colder and the region's plants began storing energy in their roots, the Indians ventured inland to find other sources of food. Eventually breaking up into small hunting groups, men set off after beaver, moose, deer, and bear, tracking the animals through the prints they left in the snow; women cleaned and prepared the meat, while tending to the campsites. In contrast to the summer, when food was in abundance, February and March often spelled privation, especially if a lack of snow made it more difficult to follow the animals. The Indians thus exploited various habitats, migrating across the landscape depending on the season of the year. As one European observer noted, "They move ... from one place to another according to the richness of the site and the season".

2 In the South, a warmer climate more conducive to agriculture allowed the Indians to combine farming with hunting and gathering to produce an even more secure subsistence diet. With the onset of warmer weather, late in February or early in March, men built fires

to clear trees and ready the ground for planting. Women then formed the soil into small hills, sowing corn and beans, while planting squash and pumpkins in trenches between the mounds. Mixing such crops together had a number of important benefits that typically led to bumper agricultural yields. As the different plants competed for sunlight and moisture, the seed stock eventually became hardier. The crop mix may also have cut down on pests, as insects found it difficult to find their favorite crop in the tangled mass of stalks. Meanwhile, bacteria found on the roots of the beans helped to replace the nitrogen that the corn sucked out of the soil, enhancing the field's fertility. But the so-called "nitrogen-fixing bacteria" were never able to add back all of the nitrogen lost, and fertility eventually declined, spurring the Indians to move on to find another area of trees to burn.

3 Southern Indians scheduled hunting and gathering around their shifting agricultural pursuits. In the spring, after burning the trees, men set off to catch fish. In the summer, the Indians along the coast moved further inland to hunt turkeys and squirrels and gather berries, returning downstream in time to harvest crops in the fall. The Indians were so attuned to the seasonal variation that characterized the forest that they often gave the months such names as "herring month" (March) or "strawberry month" (June) to describe the food they had come to expect from the landscape. Meanwhile, as the weather turned colder, nuts and acorns proliferated, attracting such game as deer and bears. As the animals fattened themselves on the food, their coats became thicker, making them inviting targets for Indians, who in the winter hunted them for meat and clothing.

4 A similar seasonal subsistence cycle based on farming and hunting and collecting prevailed further west on the Great Plains. Apart from climate, which despite the potential for drought favored agriculture, soil on the midwestern prairies was as much as a full foot deeper than the two to four inches commonly found in New England. In the valleys of the Platte, Loup, and Republican rivers in present-day Nebraska and Kansas, Pawnee men and women capitalized on the excellent soil and favorable weather conditions by first burning areas during the early spring. Women then sowed corn, beans, and squash in small plots during April and May, hoeing them periodically. Women also spent the spring gathering Indian potatoes, an abundant root crop often relied on in periods of scarcity. In July and August, the Pawnees packed the dry foods—wild and domesticated—that they harvested in the river valleys and used them to sustain themselves as they journeyed to the mixed-grass prairie west of the 98th meridian to hunt buffalo (Buffalo thrive on the grasses—blue grama, buffalo grass, and red three-awn—primarily because they find them easy to digest). In September, the Pawnees returned to the river valleys to harvest their crops, before leaving again in November for the plains to hunt buffalo. The primary goal of this system of hunting and horticulture—in existence for centuries before the coming of white settlers to the plains region—was to obtain a diversified set of food sources. It might be termed a "not-putting-all-your-eggs-in-one-basket" approach to deriving a living from the land.

5 Obviously, the Indians transformed the ecology of North America in their efforts to survive. But two points about their particular relationship with the land are worth underscoring. First, ample evidence suggests that in many instances, Native Americans exploited the landscape in a way that maintained species population and diversity. In California, for instance, Indians pruned shrubs for the purpose of basket making, but took care to do so during the dormant fall or winter period when the plant's future health would not be jeopardized. Similarly, shifting agriculture tended to mimic natural patterns in a way that modern agriculture, with its emphasis on single-crop production, does not. Second, dietary security, not the maximization of crop yields, was the most important element of Native American subsistence. At times, this decision not to stockpile food could hurt them, even if it contributed to long-term ecological balance. It was common in northern New England for Indians to go hungry and even starve during February and March (when animal populations dipped), rather than to store more food during the summer for winter use. While this failure to maximize food sources may have jeopardized Indian lives, it also helped to keep population densities relatively low. The low density, in turn, may have contributed to the overall stability of these ecosystems, preserving the future prospects of the Indians' mode of food production. North America may well have suffered from a relative lack of biological resources (at least when compared with Eurasia), but the Indians managed to see in the land a vast expanse of possibilities for ensuring food security.

(Adapted from Steinberg, 2002)

New Words and Expressions

exploit /ɪkˈsplɔɪt/ *vt.* 利用，发挥

deposit /dɪˈpɔzɪt/ *vt.* 使沉积，使沉淀，使淤积

compel /kəmˈpel/ *v.* 强迫，迫使；使必须

weir /wɪə(r)/ *n.* 堰，拦河坝

spawn /spɔːn/ *vt.* 产卵

bolster /ˈbəʊlstə(r)/ *v.* 改善，加强

privation /praɪˈveɪʃn/ *n.* 贫困，匮乏，艰难

conducive /kənˈdjuːsɪv/ *adj.* 使容易（或有可能）发生的

subsistence /səbˈsɪstəns/ *n.* 维持生计

squash /skwɔʃ/ *n.* 瓜类蔬菜

trench /trentʃ/ *n.* 沟，渠

mound /maʊnd/ *n.* 土墩；小丘，小山岗

bumper /ˈbʌmpə(r)/ *adj.* 异常大的；丰盛的

hardy /ˈhɑːdɪ/ *adj.* 耐寒的，能越冬的

tangled /ˈtæŋgld/ *adj.* 缠结的，混乱的

stalk /stɔːk/ *n.* （植物的）茎、秆

fertility /fəˈtɪlətɪ/ *n.* 富饶，丰产

attune /əˈtjuːn/ *vt.* 使协调

herring /ˈherɪŋ/ *n.* 鲱鱼

proliferate /prəˈlɪfəreɪt/ *vt.* 激增

meridian /məˈrɪdɪən/ *n.* 子午线，经线

underscore /ˌʌndəˈskɔː(r)/ *vt.* 突出显示，强调

prune /pruːn/ *v.* 修剪树枝，打杈

dormant /ˈdɔːmənt/ *adj.* 休眠的，蛰伏的；暂停活动的

jeopardize /ˈdʒepədaɪz/ *vt.* 冒某种危险；

危及,危害,损害

dietary /ˈdaɪətərɪ/ *adj.* 饮食的

stockpile /ˈstɔkpaɪl/ *vt.* 大量储备

dip /dɪp/ *v.* 下降

Exercises

Section I Knowledge Focus

Task 1 ***In North America, the Indians had to adapt to various landscapes they inhabited and manage them accordingly. Read Paras. 1-4 and complete the following diagram by filling in the blanks with words from these paragraphs.*** (***ONE WORD ONLY***)

Places	Geographic Conditions	Agricultural Situations
Far northern reaches of New England	...cold and frost... ...the 1) ________ soil deposited by the glaciers...	Agriculture was a 2) ________ and difficult venture
South	a 3) ________ climate	...more 4) ________ to agriculture
Further west on the Great Plains	... favorable weather conditions, despite the potential for drought... ...the 5) ________ soil	Climate 6) ________ agriculture

Task 2 ***To survive in North America, the Indians exploited the adapted to diversity of the landscapes they inhabited. Read Paras. 1-4 again and complete the following diagram by filling in the blanks with words from these paragraphs.*** (***ONE WORD ONLY***)

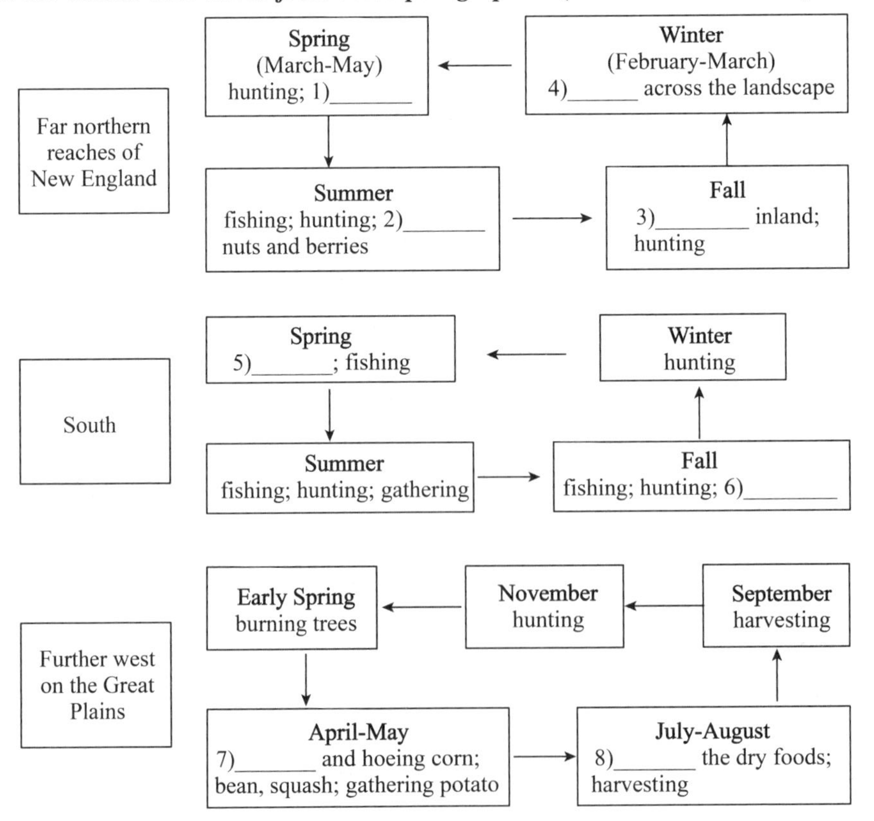

Task 3 ***Based on your understanding of the passage, decide whether the following statements are true or false. Put T for true and F for false in the blank provided before each statement.***

__________ 1) In the far northern reaches of New England, the Indians exploited various habitats, moving from one place to another depending on the season of the year.

__________ 2) In the southern part of North America, the Indians had a considerably secure subsistence diet due to a more favorable climate; thus, they didn't have to move from one place to another.

__________ 3) Further west on the Great Plains, the Pawnees capitalized on the rich soil and good weather conditions to the fullest so that they settled down on farming.

__________ 4) Indians deliberately transformed the ecology of North America in their efforts to maximize food sources in order to develop more quickly.

__________ 5) Dietary security was not the most important element of Native American subsistence.

Section II Language Focus

Task 4 ***Match the term in the left column with an explanation given in the right column and write the corresponding letter in the space provided below.***

1) mound	A. *n.* a low wall or barrier built across a river in order to control the flow of water or change its direction
2) meridian	B. *n.* a long deep hole dug in the ground, for example for carrying away water
3) weir	C. *n.* a large pile of earth or stones; a small hill
4) trench	D. *n.* one of the lines that is drawn from the North Pole to the South Pole on a map of the world

1) __________ 2) __________ 3) __________ 4) __________

Task 5 ***Complete the following sentences with appropriate words or expressions given below. Change the form where necessary.***

privation exploit dormant hardy understore

1) Most of the older organisms were nearly wiped out, although a few __________ species survived.
2) The seeds then lie __________ until the next wet year, when the desert blooms again.
3) They endured years of suffering and __________.
4) Modern technology is good enough to __________ all the sea resources.
5) The Columbian Exchange __________ the profound impact of trans – hemispheric biological exchanges on global environmental history, driving significant changes in

ecosystems, agricultural practices, and human societies worldwide.

deposit tangle prune attune proliferate

6) Elephants are poached in central and east Africa but ____________ in the south.

7) Can you tell me whether it is a good thing to ____________ an apple tree?

8) Sand was ____________ and hardened into sandstone.

9) The ____________ branches of the forest canopy affect light penetration and species diversity below.

10) He tried to ____________ himself to the Chinese way of living.

Task 6 ***Find the words in the box that have the same meaning as the underlined words or phrases in the following sentences, and write the corresponding word in the space provided.***

compel domesticated bolster conducive jeopardize

____________ 1) The fall in interest rates is starting to improve confidence.

____________ 2) The law can force fathers to make regular payments for their children.

____________ 3) More than two unexcused absences will seriously do harm to your class participation grade.

____________ 4) The newly tamed animals behaved better, were easier to control, and may have enjoyed a higher birth rate, which in turn yielded greater milk supplies.

____________ 5) Team spirit and effective communication are beneficial to employee and business performance alike.

Task 7 ***Match the words in the left column with those in the right column to form appropriate expressions. Then, complete the following sentences with one of the expressions. Change the form or add articles if necessary.***

inviting	yields
temperate	pests
crop	target
cut down on	climate

1) Although these vegetables adapt well to our ____________, they tend to crop poorly.

2) Make your own car a less ____________, to discourage thieves from trying.

3) She says the ducks ____________ such as golden apple snails while fertilizing the rice with their droppings.

4) Long before the use of machinery and fertilizer, industrious and creative farmers had already used different kinds of methods to increase ____________.

Section III Sharing Your Ideas

Task 8 ***The Indians inhabited various landscapes, and they may have suffered from a relative lack of biological resources. What did they do to survive? Did they harm the environment or take good advantage of the places they lived in? What would you do if you were the leader of an indigenous tribe?***

Active Reading 2

Warming Up

Task ***Watch a short video about Easter Island, and discuss the following questions with your partners.***

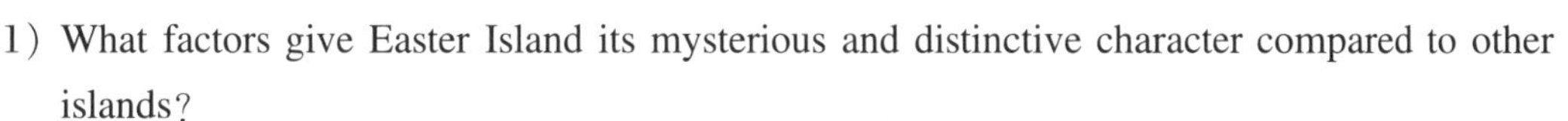

1) What factors give Easter Island its mysterious and distinctive character compared to other islands?
2) How many stone statues are present on Easter Island?
3) What explanations does the video provide for the collapse of civilization on Easter Island?
4) What else do you know about Easter Island?

Reading

Easter Island: A Lesson for Us All

1 Easter Island is one of the world's most remote places inhabited by people: 2,500 miles from the nearest continent (South America) and 1,200 miles from the nearest island (Pitcairn). At the same time, it is one of the most enchanting archaeological sites: the mysterious enormous heads dotting the island have amazed people since the discovery by Dutch sailors on Easter Sunday in 1722. The Dutch discoverers found a primitive society with about 3,000 people living in squalid reed huts or caves, engaged in almost perpetual warfare and resorting to cannibalism in a desperate attempt to supplement the meagre food supplies available on the island.

2 How could people make and transport the huge sculptures they found everywhere on the island from the quarry at the mountainside to the coast without machines, even without trees? The island was completely treeless at the time of discovery. The story of the Easter islanders is tragic, but at the same time, a good lesson for all of us. They had a highly developed civilisation for about 600 years, but neglected the environmental effect of their lifestyle and ended in catastrophe. They could not escape the island anymore, because every single tree had been cut.

3 Modern research has revealed a lot about their life during their golden age-and about the causes of the catastrophe.

The History of the Islanders

4 The original Polynesians came from south-east Asia. They made long voyages in double canoes, joined together by a broad central platform to transport and shelter people, plants, animals and food. When the first people found Easter Island, they discovered a world with few resources. The island was volcanic in origin, but its three volcanoes had been extinct for many centuries before the Polynesian settlers arrived. Because of its remoteness the island had only a few species of plants and animals. There were thirty indigenous plants, no mammals, but many seabirds.

5 The people who arrived in the fifth century probably numbered no more than twenty or thirty at most. The settlers on Easter Island brought only chickens and rats with them and because the climate was too severe for many plants grown elsewhere in Polynesia, they were restricted to a diet based mainly on sweet potatoes and chickens. The only advantage of this monotonous, though nutritionally adequate, diet was that cultivation of the sweet potato was not very demanding and left plenty of time for other activities. People had time for cultural development. The result was the creation of the most advanced of all the Polynesian societies and one of the most complex in the world. The Easter Islanders engaged in elaborate rituals and monument construction.

Statues on the Mountainside

6 The crucial centres of ceremonial activity were the *ahu*. Over 300 of these platforms were constructed on the island, mainly near the coast. A number of these *ahu* have sophisticated astronomical alignments, towards one of the solstices or the equinox. Rock paintings and scripts on wooden panels have also been found. At each site between one and fifteen of the huge stone statues survive today as a unique memorial to the vanished Easter Island society. These statues took up immense amounts of peasant labour. The most challenging problem was to transport the statues, each some twenty feet in length and weighing several tens of tons, across the island and then erect them on top of the *ahu*. Lacking any draught animals they had to rely on human power to drag the statues across the island using tree trunks as rollers.

7 The only way this could have been accomplished was by large numbers of people guiding and sliding them along a flexible track made of tree trunks spread on the ground between the quarry and the *ahu*. Enormous quantities of timber were required.

8 The population of the island grew steadily from the original small group to about 7,000 at its peak in 1550. By the sixteenth century hundreds of *ahu* had been constructed and with them over 600 of the huge stone statues.

9 Then, when the society was at its peak, it suddenly collapsed leaving over half the statues only partially completed around Rano Raraku quarry. The cause of the collapse and the key

to understanding the "mysteries" of Easter Island was massive environmental degradation brought on by deforestation of the whole island.

The Science: Ecological and Archaeological

10 Recent scientific work, involving the analysis of pollen types, has shown that at the time of the initial settlement Easter Island had a dense vegetation cover including extensive woods. As the population slowly increased, trees have been cut down to provide space for agriculture, fuel for heating and cooking, construction material for household goods, pole-and-thatch houses and canoes for fishing. The most demanding requirement was the need to move the large number of enormously heavy statues to ceremonial sites around the island. As a result, by 1600, the island was almost completely deforested and statue erection was brought to a halt leaving many statues stranded at the quarry.

11 The deforestation of the island meant not only the end of the elaborate social and ceremonial life; it also had other drastic effects on everyday life for the population generally.

12 Archaeological research shows that from 1500 the shortage of trees was forcing many people to abandon building houses from timber and live in caves. They resorted to stone shelters dug into the hillsides or flimsy reed huts cut from the vegetation that grew round the edges of the crater lakes. Canoes could no longer be built and only reed boats incapable of long voyages could be made. Fishing was also more difficult because nets had previously been made from the paper mulberry tree (which could also be made into cloth) and that was no longer available. No new trees could grow, because the rats, imported for food, ate the fruits and seeds. In archaeological sites, nuts and seeds were found, all visibly opened by rats.

13 Removal of the tree cover also badly affected the soil of the island. Increased exposure caused soil erosion and the leaching out of essential nutrients. As a result, crop yields declined. The only source of food on the island unaffected by these problems was the chickens. The society went into decline and regressed to ever more primitive conditions. Without trees, and therefore without canoes, the islanders were trapped in their remote home, unable to escape the consequences of their self-inflicted, environmental collapse. There were increasing conflicts over diminishing resources resulting in a state of almost permanent warfare. Slavery became common and as the amount of protein available fell the population turned to cannibalism.

14 The magnificent stone statues, too massive to destroy, were pulled down. The first Europeans found only a few still standing and all had been toppled by the 1830s. When the Europeans asked how the statues had been moved from the quarry, the primitive islanders could no longer remember what their ancestors had achieved and could only say

that the huge figures had "walked" across the island. The Europeans, seeing a treeless landscape, were equally mystified. They speculated the most fantastic explanations.

The Lesson of Easter Island

15 Against great odds the islanders had painstakingly constructed, over several centuries, one of the most advanced societies of its type in the world. For a thousand years they sustained a way of life not only to survive but to flourish. It was in many ways a triumph of human ingenuity and an apparent victory over a difficult environment. But in the end the increasing numbers and cultural ambitions of the islanders proved too great for the limited resources available to them. When the environment was ruined by the pressure, the society very quickly collapsed with it, leading to a state of near barbarism.

16 The Easter Islanders, aware that they were almost completely isolated from the rest of the world, must surely have realized that their very existence depended on the limited resources of a small island. They must have seen what was happening to the forests. Yet they were unable to devise a system that allowed them to find a balance with their environment. Instead, vital resources were steadily consumed until finally none were left. Indeed, at the time when the coming catastrophe must have become starkly apparent, more and more statues were carved and moved across the island. The fact that so many were left unfinished or stranded near the quarry suggests that no account was taken of how few trees were left on the island.

17 The fate of Easter Island can be a lesson for the modern world too. Like Easter Island, Earth has only limited resources to support human society and all its demands. Like the islanders, the human population of the earth has no practical means of escape. How has the environment of the world shaped human history and how have people shaped and altered the world in which they live? Have other societies fallen into the same trap as the islanders? For the last few millennia humans have succeeded in obtaining more food and extracting more resources for increasing numbers of people and increasingly complex and technologically advanced societies. But are we now any more successful than the islanders in finding a way of life that does not fatally deplete the resources that are available to us or are we too busy irreversibly damaging our life support system?

(Adapted from Sustainable Footprint, 2022)

New Words and Expressions

enchanting **/ɪnˈtʃɑːntɪŋ/** *adj.* 迷人的,令人陶醉的

archaeological **/ˌɑːkɪəˈlɒdʒɪk(ə)l/** *adj.* 考古学的,考古的

primitive **/ˈprɪmətɪv/** *adj.* 原始的,远古的

squalid **/ˈskwɒlɪd/** *adj.* (处所及生活环境)肮脏的,邋遢的

cannibalism **/ˈkænɪbəlɪzəm/** *n.* 食人,同类

相食
meagre /ˈmiːɡə(r)/ *adj.* 少量且劣质的
quarry /ˈkwɔrɪ/ *n.* 采石场
catastrophe /kəˈtæstrəfɪ/ *n.* 灾难,灾祸,横祸
volcanic /vɔlˈkænɪk/ *adj.* 火山的,火山引起的,火山产生的
volcano /vɔlˈkeɪnəʊ/ *n.* 火山
extinct /ɪkˈstɪŋkt/ *adj.* (火山)死的
indigenous /ɪnˈdɪdʒənəs/ *adj.* 本地的,当地的,土生土长的
monotonous /məˈnɔtənəs/ *adj.* 单调乏味的
astronomical /ˌæstrəˈnɔmɪkl/ *adj.* 天文学的,天文的
alignment /əˈlaɪnmənt/ *n.* 排成直线
solstice /ˈsɔlstɪs/ *n.* 至(点);(夏或冬)至
equinox /ˈiːkwɪnɔks/ *n.* 二分时刻;昼夜平分时;春分,秋分
degradation /ˌdegrəˈdeɪʃn/ *n.* 毁坏,恶化
deforestation /ˌdiːˌfɔrɪˈsteɪʃn/ *n.* 毁林,滥伐森林,烧林
pollen /ˈpɔlən/ *n.* 花粉
strand /strænd/ *v. t* 使滞留
flimsy /ˈflɪmzɪ/ *adj.* 脆弱的;劣质的
leach /liːtʃ/ *v.* 过滤
regress /rɪˈgres/ *vi.* 倒退,退化
self-inflicted /ˌself ɪnˈflɪktɪd/ *adj.* 自己造成的
topple /ˈtɔpl/ *v.* 倒塌,倒下
mystify /ˈmɪstɪfaɪ/ *vt.* 迷惑,使迷惑不解,使糊涂
ingenuity /ˌɪndʒəˈnjuːətɪ/ *n.* 聪明才智;心灵手巧
barbarism /ˈbaːbərɪzəm/ *n.* 野蛮,未开化,不文明
starkly /ˈstaːklɪ/ *adv.* 明显地,毫无掩饰地
millennia /mɪˈlenɪə/ *n.* 一千年
deplete /dɪˈpliːt/ *vt.* 大量减少;耗尽,使枯竭
irreversibly /ˌɪrɪˈvɜːsəblɪ/ *adv.* 不可逆地

Exercises

Section I Knowledge Focus

Task 1 ***Read Paras. 1-10 and complete the following diagram by filling in the blanks with words from these paragraphs.*** (***ONE WORD ONLY***)

Time	5th century	By the 16th century	1722
People	Polynesians	Easter Islanders	Dutch sailors Easter Islanders
Population	less than 20-30	6)__________	3,000
Natural conditions	It was 1)__________ in origin. There were thirty indigenous plants, no 2)__________, but many seabirds. It had a dense 3)__________ cover including extensive woods.	The island was almost completely 7)__________.	treeless

（续）

Time	5th century	By the 16th century	1722
Activities	People had time for cultural development. The Easter Islanders engaged in elaborate 4) ________ and 5) ________ construction.	Trees have been cut down to provide space for 8) ________, fuel for heating and cooking, construction material for household goods, houses and canoes for 9) ______. Hundreds of *ahu* had been constructed and with them over 600 of the huge stone 10) ________.	Living in squalid reed huts or caves, people engaged in almost perpetual warfare and resorting to 11) ________ in a desperate attempt to supplement the 12) ________ food supplies available on the island.

Task 2 ***The deforestation of Easter island meant not only the end of the elaborate social and ceremonial life; it also had other drastic effects on everyday life for the population as a whole. Read the whole passage and complete the following diagram by filling in the blanks with words from the passage.*** （***ONE WORD ONLY***）

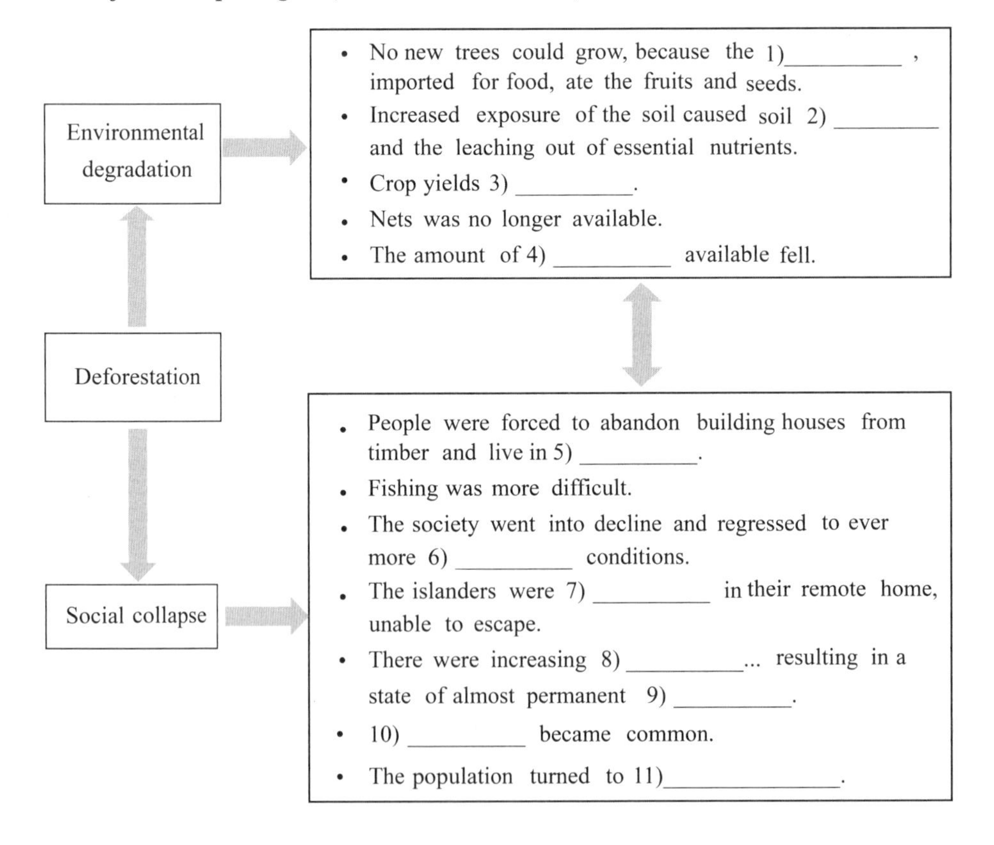

Section II Language Focus

Task 3 ***Match the term in the left column with an explanation given in the right column and write the corresponding letter in the space provided below.***

1) cannibalism	A. *n.* arrangement in a straight line
2) catastrophe	B. *n.* eating the flesh of other people
3) alignment	C. *n.* a sudden event that causes many people to suffer
4) barbarism	D. *n.* the quality of being clever, original and inventive
5) ingenuity	E. *n.* absence of culture and civilization

1)__________ 2)__________ 3)__________ 4)__________ 5)__________

Task 4 ***Complete the following sentences with appropriate words or expressions given below. Change the form where necessary.***

enchanting indigenous regress degradation topple

1) This raises unique challenges for Arctic wildlife and __________ people who depend on Arctic ecosystems to survive.
2) We were attacked relentlessly by winds that tried to __________ it over.
3) The main attraction of Beijing resides in its __________ scenery.
4) Some children may __________ to an earlier stage of development.
5) Species extinction, deforestation, damming of rivers, occurrence of floods, the depletion of ozone, the __________ of ocean systems and many other areas are all experiencing acceleration.

squalid meagre flimsy drastic strand

6) Most advertisements are just too __________ in informative content and too rich in emotional suggestive detail to be read literally.
7) We're really just now beginning to understand how quickly __________ climate change can take place.
8) The ship was __________ on a sandbank.
9) It's the money that spurs these fishermen to risk a long ocean journey in their __________ boats.
10) The early industrial cities were __________ and unhealthy places.

Task 5 ***Translate the English expressions in Column A into Chinese, and the Chinese ones in Column B into English.***

A	B
1) archaeological site	8) 红薯
2) dense vegetation cover	9) 农作物产量

续表

A	B
3) elaborate rituals	10) 黄金时代
4) at the peak	11) 役畜
5) consume resources	12) 土壤侵蚀
6) against odds	13) 家庭用品
7) end in catastrophe	14) 原始社会

Task 6 ***Translate the following English paragraph into Chinese.***

The fate of Easter Island can be a lesson for the modern world too. Like Easter Island the Earth has only limited resources to support human society and all its demands. Like the islanders, the human population of the earth has no practical means of escape. How has the environment of the world shaped human history and how have people shaped and altered the world in which they live? Have other societies fallen into the same trap as the islanders? For the last few millennia humans have succeeded in obtaining more food and extracting more resources for increasing numbers of people and increasingly complex and technologically advanced societies. But are we now any more successful than the islanders in finding a way of life that does not fatally deplete the resources that are available to us or are we too busy irreversibly damaging our life support system?

Section III Sharing Your Ideas

Since you have learned about the two civilizations from The civilizations in North America and Easter Island were two different types of human-environment interaction: The former being complete dependence on the environment, and the latter involving environmental modification to a disruptive, even catastrophic extent. Anyhow, both civilizations suffered greatly from their respective types of human-environment interaction and gradually diminished, or even collapsed. Of course, aside from their fragile relationship with nature, there are many other contributing factors. However, one thing is crystal-clear—both serve as important examples, urging us to develop a more sustainable way to coexist with nature.

Task 7 ***Not only has the human population, but also the energy consumption, has growing exponentially. Compare your daily energy consumption with that of your grandparents when they were young (ask them if possible). What steps can you take to reduce energy usage?***

Task 8 ***You could conduct a study to determine whether the story of Easter Island is unique in world history. Have other civilizations collapsed as a result of unhealthy human-environment interactions? What changes should be made in our society (or lifestyle) to prevent such a collapse?***

Part Three China's Environmental Story

Active Reading 3

Warming Up

Task ***The wooly mammoth and the saber-toothed tiger are enduring symbols of the Ice Age, serving as testaments to the era's unique biological diversity. Discuss with your partners what you know about them.***

Reading

China's Natural Environment and Early Human Settlement

1 China's natural environment was exceptionally diverse, providing large numbers of niches for a wide variety of plants and animals to exploit. In that environment, climatic and other changes that ended the last Ice Age also created conditions in which people could domesticate plants and begin farming. This happened as early as 9,500 to 8,800 years ago in the Yangtze River Valley, where some people began to cultivate rice, and by eight thousand years ago in northern China, where dry farming based on millet emerged. Over several thousand years, as agriculture provided food for growing populations, villages came under the control of rulers, and the first states emerged.

2 The reason for China's extraordinary biodiversity, of course, is that it has such a variety of geography, climate, and soils. It spans 50 degrees of latitude and 62 degrees of longitude, and ranges in altitude from nearly 1,000 feet below sea level (the Turpan Depression in modern Xinjiang in northwest China) to more than 29,000 feet on the peak of Qomolungma. From the tropics of Hainan Island and southern China, with annual mean temperatures in the 70s (in degrees Fahrenheit) and annual rainfall of 80 inches, to glacial conditions on the Tibetan Plateau, with annual mean temperatures around 40 degrees and little rain, to the coniferous forests of the northeast, China has a vast range of climates. It also has most of the soil types found in the world.

3 However, because of the rapid falloff in the amount of monsoonal rain reaching into the interior, according to Nicholas Menzies, "the forests of China are largely confined to the eastern half of the country. Grasslands and scrub predominate in the arid west, except in mountainous areas where sufficient water is derived from winter snowfall to support coniferous forests..." The historical geographer Wen Huanjan has identified five forest

zones (not including the western grasslands and desert) which combine the biological factor of the dominant plant communities with the human factor of the history of the forest. These are the boreal forest (of the northeast), the temperate forest of northern China, the subtropical forest of central and southwestern China, and the tropical forest of southern China.

4 Forests are not merely stands of trees but, as noted above, "communities" that scientists now recognize as ecosystems, with varieties of organisms ranging from bacteria in the soil to the large mammals—often carnivores—at the top of the food chain, all interdependent and interacting with each other as well as with soil, water, and solar energy. The more species of plants and animals interacting in any given ecosystem, the more biologically diverse and, hence, healthier the ecosystem is.

5 In summary, the physical space we now call China spanned an exceptional number of ecosystems, from tropical rain forests and coral seas in the south to the highest mountains on earth in the west, to grassland and desert in the northwest and an unbroken forest stretching from the south to the northern grasslands. In light of the vast numbers of ecosystems and ecological niches involved, as well as China never having had glaciers, it should not be too surprising that China was home to an immense number of different animal and plant species—China was species-rich. Such an environment was also hospitable to another species of animal searching for sustenance from its surroundings—humans.

6 By the time anatomically modern humans (*Homo sapiens*) migrated out of Africa and through Southeast Asia into East Asia, fire-and tool-using hominids were already there. These early humans disappeared from Asia about fifty thousand years ago (in Europe between thirty-three thousand and twenty-four thousand years ago), possibly driven out of existence by competition for food from the more technologically advanced modern humans. The total population of the earlier hominids probably never exceeded ten thousand individuals, so their environmental impact was small. So, too, was the impact of early human hunter-gatherers, although some paleobiologists suspect that their hunting prowess may have been responsible for the disappearance of several large animals such as the wooly mammoth and saber-toothed tiger. Whatever the case, it is certain that the major change in terms of the human relationship to their environment came with the emergence of agriculture, beginning about 9,500 years ago, not just in China but in at least four other parts of the world as well (the Fertile Crescent of ancient Mesopotamia, in what is now Mexico and the Andes in Mesoamerica, and on what is now the eastern coast of the United States, about 4,500 years ago, and probably also in west Africa and New Guinea). Thus, China was one of the few places on earth where agriculture not only developed early but also became a mainstay of human society, and agriculture is thus central to the story of

China's environmental history. People might think of it this way: hunters and gatherers depended on maintaining their environment more or less unchanged to ensure the supply of the game and fruits and nuts that sustained their populations. They probably did use fire in the forest to clear out the underbrush and to allow fresh grasses to grow so as to attract deer and kill them more easily. But the forest remained. Settled farming, on the other hand, requires that clearings in the forest be made and, when accompanied by the farmers' settlement in permanent villages, that those clearings be maintained. Farming increased the food supply by making its production more certain, and thus the human population increased in number as well, creating a need for more farmland. The technologies of farming and settled agriculture are thus significant ways in which humans have interacted with and transformed their environment.

(Adapted from Marks,2011)

New Words and Expressions

niche /niːtʃ/ *n.* 生态位(一个生物所占生境的最小单位)

millet /ˈmɪlɪt/ *n.* 小米,粟

latitude /ˈlætɪtjuːd/ *n.* 纬度

longitude /ˈlɒŋgɪtjuːd/ *n.* 经度

altitude /ˈæltɪtjuːd/ *n.* 海拔

glacial /ˈgleɪʃ(ə)l/ *adj.* 冰川造成的,由冰河形成的;冰河的,冰川的

coniferous /kəˈnɪfərəs/ *adj.* 针叶的

falloff /ˈfɔːlɔf/ *n.* 下降,减少;减退

monsoonal /mɔnˈsuːnl/ *adj.* 季风的

scrub /skrʌb/ *n.* 硬叶灌丛带,低矮灌木丛林地

predominate /prɪˈdɔmɪneɪt/ *v.* (数量上)占优势

arid /ˈærɪd/ *adj.* 干旱的,干燥的

geographer /dʒɪˈɔgrəfə(r)/ *n.* 地理学研究者,地理学家

exploitation /ˌeksplɔɪˈteɪʃ(ə)n/ *n.* 利用,开发,开采

boreal /ˈbɔːrɪəl/ *adj.* 北方的;北风的

carnivore /ˈkɑːnɪvɔː(r)/ *n.* 食肉动物

interdependent /ˌɪntədɪˈpendənt/ *adj.* (各部分)相互依存的,相互依赖的

sustenance /ˈsʌstənəns/ *n.* 食物;营养,养料

anatomically /ˌænəˈtɔmɪklɪ/ *adv.* 解剖学上

hominid /ˈhɔmɪnɪd/ *adj.* 人类及其祖先的

paleobiologist /ˌpælɪəʊbaɪˈɔlədʒɪst/ *n.* 古生物学家

prowess /ˈpraʊəs/ *n.* 非凡的技能,高超的技艺

mainstay /ˈmeɪnsteɪ/ *n.* 支柱,中流砥柱

underbrush /ˈʌndəbrʌʃ/ *n.* (森林里树木下的)下层灌丛

Exercises

Section I Understanding the Text

Task 1 ***Discuss the following questions in small groups.***

1) When and where did people begin to domesticate plants and engage in farming in China, according to the passage?

2) Why does China have such an extraordinary biodiversity?

3) How many forest zones can be identified in China according to historical geographer Wen Huan-Jan, and what are they?

4) What factors make China hospitable to human beings?

5) When and where did the major changes occur in terms of the human relationship with the environment?

6) What are the significant ways in which humans have interacted with and transformed the environment according to the passage?

Section II Developing Critical Thinking

To protect our environment, it is crucial that every individual acts as a protector, builder and beneficiary, with no one standing idly by, remaining indifferent, or merely offering criticism.

Task 2 ***Suppose you are in a city where people pay less attention to nature and more to economic development. To support eco-environmental protection and to change the common view on sustainable development, you are going to carry out a debate about the proposition "No durable progress without strong bottom-up activity".***

Your goal is to convince people in your city that every individual must act in their everyday life. To make your viewpoint more persuasive and acceptable, consider following the instruction below.

How to argue?

There is no set model of organization for argumentation. Below are some common patterns.

Patterns	Description
Pattern 1	· **Provide** your first major sub-argument and supporting evidence. · **Provide** your second major sub-argument and supporting evidence. … · **Refute** your opponents' first point. · **Refute** your opponents' second point. …
Pattern 2	· **Refute** your opponents' first point. · **Refute** your opponents' second point. … · **Provide** your first major sub-argument and supporting evidence. · **Provide** your second major sub-argument and supporting evidence. …

(续)

Patterns	Description
Pattern 3	· **Provide** your first major sub-argument and supporting evidence, which also refutes one of your opponents' points. · **Provide** your second major sub-argument and supporting evidence, which also refutes one of your opponents' points. …

There are two key issues in an argumentation: evidence and logical reasoning.

Evidence: To back up your argumentation, you need to provide strong evidence. No matter what evidence you may use, it must be accurate, up-to-date, sufficient and relevant to your viewpoint.

Logical reasoning: Logical reasoning takes two forms, deductive and inductive respectively. And for your argumentation, you are supposed to avoid some common fallacies.

1) Rush to a conclusion on the basis of insufficient or biased evidence.
2) Assume that if "A" occurred after "B", "B" must be the cause of "A".
3) Make a circular statement, i. e. to repeat the argument instead of verifying it.
4) Make a false analogy, i. e. to wrongly assume that two things are similar in some ways, they must be alike in all ways.
5) Avoid opponents' arguments rather than addressing them.

(Adapted from Ji, 2013)

UNIT 3

Agrarian Civilization and Environment

Part One　Preparation

Unit Preview

For a long period, people were hunter-gatherers, using simple tools to hunt animals or gather wild plants for sustenance until the shift to farming around 12,000 years ago. This transition marked the beginning of agrarian society, where arable land became the primary source of wealth, focusing on agriculture. People at that time typically settled along the flood plains of rivers, since the rivers flooded periodically, depositing nutrient-rich silt that made the soil highly suitable for growing crops.

As the Bronze Age dawned, a remarkable acceleration in agricultural development occurred across a number of civilizations, including Sumer in Mesopotamia, ancient Egypt, the Indus Valley civilization on the Indian subcontinent, and ancient China. However, the practice of agriculture frequently disrupted natural ecosystems. The introduction of irrigation systems often had profound environmental impacts, sometimes leading to disastrous outcomes. Such environmental degradation was often a precursor to the decline of civilizations.

Among these ancient civilizations, China, as the world's oldest continuous civilization, not only exemplifies a sustainable approach to land management and natural resource utilization, but also demonstrates a commitment to preserving and harnessing nature. And by doing this, the Chinese civilization has also managed to keep its culture alive and contribute to the broader advancement and evolution of global civilization.

Learning Objectives

Upon completion of this unit, you will be able to:

- analyze the characteristics and challenges of early agrarian societies.
- evaluate the impact of the natural environment on the development of agrarian societies.
- explain the relationship between agricultural practices and the rise or collapse of civilizations.

Part Two Global Perspectives

Active Reading 1

Warming Up

视听资源

Task 1 ***Discuss the following questions with your partners.***

1) What is the first written language in human history?

2) What else do you know about it?

Task 2 ***Watch a short video clip. Then discuss the following questions with your partners.***

1) How did the ancient Sumerians live and survive in the early days?

2) What difficulties did the Sumerians face when they started farming?

3) What did the Sumerians invent for boosting farm productivity?

Reading

Agriculture and the Decline of Sumer

1 When, in 1936, one of the excavators of the earliest cities of Sumer, Leonard Woolley, wrote a book about his work entitled *Ur of the Chaldees* he was puzzled by the desolate, largely treeless landscape of contemporary southern Mesopotamia, similar to that imagined by Shelley.

Only to those who have seen the Mesopotamian desert will the evocation of the ancient world seem well-nigh incredible, so complete is the contrast between past and present ... it is yet more difficult to realise, that the blank waste ever blossomed, bore fruit for the sustainance [sic] of a busy world. Why, if Ur was an empire's capital, if Sumer was once a vast granary, has the population dwindled to nothing, the very soil lost its virtue?

2 The answer to Woolley's question is that the Sumerians themselves destroyed the world they had created so painstakingly out of the difficult environment of southern Mesopotamia.

3 The valley of the twin rivers, the Tigris and Euphrates, posed major problems for any society, especially in the south. The rivers were at their highest in the spring following the melting of the winter snows near their sources and at their lowest between August and October, the time when the newly planted crops needed the most water. In the north of Mesopotamia the problem was eased by the late autumn and winter rains but rainfall was very low and often non-existent further south. This meant that in the Sumerian region

water storage and irrigation were essential if crops were to be grown. At first the advantages outweighed the disadvantages but slowly a series of major problems became apparent. In summer, temperatures were high, often up to about 40℃, which increased evaporation from the surface and as a consequence the amount of salt in the soil. Water retention in the deeper layers of the soil and hence the risk of waterlogging was increased by two factors. The soil itself had very low permeability.

4 This was exacerbated by the slow rate of drainage caused by the very flat land, itself made worse by the amount of silt coming down in the rivers, probably caused by deforestation in the highlands, which added about 150 cm of silt every millennium and caused the delta of the two rivers to extend by about 25 kilometres a millennium. As the land became more waterlogged and the water table rose, more salt was brought to the surface where the high evaporation rates produced a thick layer. Modern agricultural knowledge suggests that the only way to avoid the worst of these problems is to leave the land fallow and unwatered for long periods to allow the level of the water table to fall. The internal pressures within Sumerian society made this impossible and brought about disaster. The limited amount of land that could be irrigated, rising population, the need to feed more bureaucrats and soldiers and the mounting competition between the city states all increased the pressure to intensify the agricultural system. The overwhelming requirement to grow more food meant that it was impossible to leave land fallow for long periods. Short-term demands outweighed any considerations of the need for long-term stability and the maintenance of a sustainable agricultural system.

5 About 3000 BCE Sumerian society became the first literate society in the world. The detailed administrative records kept by the temples of the city states provide a record of the changes in the agricultural system and an insight into the development of major problems. About 3500 BCE roughly equal amounts of wheat and barley were grown in southern Mesopotamia. But wheat can only tolerate a salt level of half of one per cent in the soil whereas barley can still grow in twice this amount. The increasing salinisation of the soil can be deduced from the declining amount of wheat cultivated and its replacement by the more salt-tolerant barley. By 2500 BCE wheat had fallen to only 15 per cent of the crop; by 2100 BCE Ur had abandoned wheat production and overall it had declined to just 2 per cent of the crops grown in the Sumerian region.

6 Even more important than the replacement of wheat by barley was the declining yield of crops throughout the region. In the earliest phases of Sumerian society when areas went out of production because of salinisation they were replaced by newly cultivated fields. Rising population, and the demand for a greater food surplus to maintain the army as warfare became more frequent, reinforced the demand for new land.

7 But the amount of new land that could be cultivated, even with the more extensive and

complex irrigation works that were becoming common, was limited. Until about 2400 BCE crop yields remained high, in some areas at least as high as in medieval Europe and possibly even higher. Then, as the limit of cultivatable land was reached and salinisation took an increasing toll, the food surplus began to fall rapidly. Crop yields fell by 40 per cent between 2400 and 2100 BCE and by two-thirds by 1700 BCE. From 2000 BCE there are contemporaneous reports that "the earth turned white", a clear reference to the drastic impact of salinisation. The consequences for a society so dependent on a food surplus were predictable. The bureaucracy, and perhaps even more importantly the army, could not be maintained. As the size of the army fell the state became very vulnerable to external conquest. What is remarkable is the way that the political history of Sumer and its city states so closely follows the decline of the agricultural base. The independent city states survived until 2370 BCE when the first external conqueror of the region-Sargon of Akkad-established the Akkadian empire. That conquest took place following the first serious decline in crop yields resulting from widespread salinisation. For the next six hundred years the region saw the Akkadian empire conquered by the Guti nomads from the Zagros mountains, a brief revival of the region under the Third Dynasty of Ur between 2113 and 2000 BCE, its collapse under pressure from the Elamites in the west and the Amorites in the east, and in about 1800 BCE the conquest of the area by the Babylonian kingdom centred on northern Mesopotamia. Throughout this period, from the end of the once flourishing and powerful city states to the Babylonian conquest, crop yields continued to fall, making it very difficult to sustain a viable state. By 1800 BCE, when yields were only about a third of the level obtained during the Early Dynastic period, the agricultural base of Sumer had effectively collapsed and the focus of Mesopotamian society shifted permanently to the north, where a succession of imperial states controlled the region, and Sumer declined into insignificance as an underpopulated, impoverished backwater of an empire.

8 The artificial system that was the foundation of Sumerian civilisation was very fragile and in the end brought about its downfall. The later history of the region reinforces the point that all human interventions tend to degrade ecosystems and shows how easy it is to tip the balance towards destruction. It also suggests that it is very difficult to redress the balance or reverse the process once it has started. Centuries later, when the city states of Sumer were no longer even a memory, the same processes were at work elsewhere in Mesopotamia. Between 1300 and 900 BCE there was an agricultural collapse in the central area following salinisation as a result of too much irrigation.

9 Around Baghdad in the seventh and eighth centuries CE, both before and after the Arab conquest, the area was flourishing with high crop yields from irrigated fields supporting a wealthy and sophisticated society. But the same pressures seem to have been apparent as in

Sumer over 3,000 years earlier. To boost food production four major new irrigation canals were dug between the Tigris and Euphrates, which, in turn, led to waterlogging, a rapidly rising water table and salinisation. At this time the population of Mesopotamia was probably about one-and-a-half million but the agricultural collapse brought about through intensive irrigation and the Mongol conquest in the thirteenth century caused a massive decline in population to about 150,000 by 1,500 and brought about the end of the sophisticated society that had survived in the area for centuries.

(Adapted from Ponting,2007)

New Words and Expressions

excavator /ˈekskəveɪtə(r)/ *n.* 开凿者

desolate /ˈdesələt/ *adj.* 荒无人烟的,荒凉的

evocation /ˌiːvəʊˈkeɪʃn/ *n.* 唤起,召唤

granary /ˈgrænərɪ/ *n.* 粮仓,谷仓

painstakingly /ˈpeɪnzteɪkɪŋlɪ/ *adv.* 煞费苦心地;费力地

pose /pəʊz/ *vt.* 造成,引起(危险、威胁、问题等)

evaporation /ɪˌvæpəˈreɪʃn/ *n.* 蒸发

retention /rɪˈtenʃ(ə)n/ *n.* 保留,保持;保存,存放

permeability /ˌpɜːmɪəˈbɪlətɪ/ *n.* 渗透性

exacerbate /ɪgˈzæsəbeɪt/ *vt.* 使恶化,使加剧

silt /sɪlt/ *n.* 淤泥,泥沙

waterlogged /ˈwɔːtəlɔgd/ *adj.* 涝的,浸满水的

fallow /ˈfæləʊ/ *adj.* 休耕的

salinisation /ˌsælɪnaɪˈzeɪʃən/ *n.* 盐化,盐渍化

deduce /dɪˈdjuːs/ *vt.* 推断,推理

salt-tolerant /sɔːltˈtɔlərənt/ *adj.* 耐盐的

revival /rɪˈvaɪv(ə)l/ *n.* 复兴,复苏

succession /səkˈseʃn/ *n.* 继任,继承

impoverished /ɪmˈpɔvərɪʃt/ *adj.* 贫困的,赤贫的

intervention /ˌɪntəˈvenʃ(ə)n/ *n.* 干预,介入

Exercises

Section I Knowledge Focus

Task 1 ***Read Paras. 1-6 and complete the following diagram by filling in the blanks with words from these paragraphs. Change the form where necessary.*** **(*ONE WORD ONLY*)**

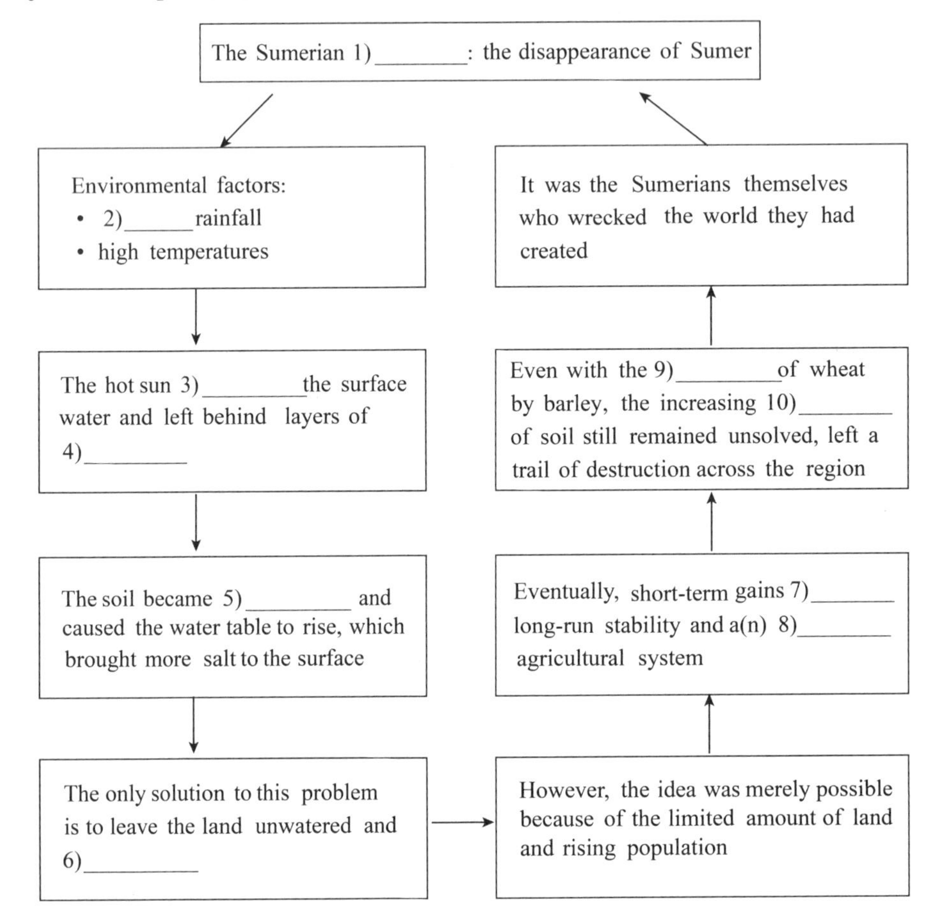

Task 2 ***Read Paras. 7-8 and complete the following table by filling in the blanks with words from these paragraphs.*** **(*ONE WORD ONLY*)**

Timeline	Agricultural Situations	Consequences
2400 BCE	High crop 1) ____________, in some areas even higher than before.	
2400—2100 BCE	As more lands suffered from soil salinity, the shortage of 2) ____________ land became a daunting problem.	The Sumerian region saw a dramatic fall of food 3) ____________.
From 2000 BCE	Early records in Sumerian writing report that "the earth turned white".	The very basis of societies became rather 4) ____________ to invaders.

(续)

Timeline	Agricultural Situations	Consequences
By 1800 BCE	The harvest of crops declined by two-thirds compared with the Early Dynastic Period.	Sumer turned into an underpopulated and 5) ___________ empire.

Task 3 ***Based on your understanding of Paras. 8-9, decide whether the following statements are true or false. Put T for true and F for false in the blank provided before each statement.***

___________ 1) Poorly functioning ecosystems caused by human activities can have irreversible effects and will undermine human welfare, as evidenced by the unfortunate subsequent history of the area.

___________ 2) The tragic story of the Sumerian cities recurred in other regions of Mesopotamia.

___________ 3) A well-established and prosperous society around Baghdad was protected from waterlogging and salinization by the advanced irrigation canals constructed between the two rivers.

___________ 4) The construction of canals between the Tigris and Euphrates eventually contributed to the decline of Mesopotamian civilization.

___________ 5) The breakdown of agriculture in Mesopotamia was likely due to multiple factors, including the extensive use of irrigation systems and the Mongol conquest.

Section II Language Focus

Task 4 ***Match the term in the left column with an explanation given in the right column and write the corresponding letter in the space provided below.***

1) irrigation	A. *n.* the saturation of ground with water
2) waterlogging	B. *n.* a build-up of salt in soil
3) silt	C. *n.* supplying dry land with water by means of ditches
4) salinisation	D. *n.* mud or clay or small rocks deposited by a river or lake

1) ___________ 2) ___________ 3) ___________ 4) ___________

Task 5 ***Complete the following sentences with appropriate words or expressions given below. Change the form where necessary.***

retention dwindle pose desolate exacerbate

1) The rhinoceros are considered endangered with ___________ wild populations.

2) Drought had ___________ the farming town.

3) Several studies have suggested that acute exercise in a hot environment could ___________ the concentration of oxidative stress.

4) Storm water __________ systems collected gray water that could be used to flush toilets and irrigate landscaping.

5) Rising unemployment is __________ problems for the administration.

delta deduce painstakingly tolerate flourish

6) As recently reviewed, rice generally can __________ flooded soil when flooding is relatively shallow.

7) The Niger __________ is rich in oil but is also an area of great poverty, creating a volatile situation.

8) The increasing erosion of the soil can be __________ from the declining amount of production.

9) Mayan civilization __________ throughout Central America during the Classic period.

10) The work had been done with __________ attention to detail.

Task 6 ***Find the words in the box that have the same meaning as the underlined words or phrases in the following sentences, and write the corresponding word in the space provided.***

intervention storage degradation cultivatable impoverished

__________ 1) The problem can be alleviated by building more water <u>retention</u> infrastructure, but that costs money.

__________ 2) Shifting cultivation has been blamed for large-scale deforestation and forest <u>degeneration</u> in the tropical region.

__________ 3) Any outside or human <u>interference</u> can drastically affect the ecosystem and wildlife.

__________ 4) Just a generation ago, the nation was overwhelmingly rural and <u>poverty-stricken</u>.

__________ 5) By 1995, only 27 percent of the <u>arable</u> land has been utilized in that region.

Task 7 ***Match the words in the left column with those in the right column to form appropriate expressions. Then complete the following sentences with one of the expressions. Change the form or add articles where necessary.***

exacerbate	salinisation
pose	fallow
lie	irrigation
undergo	deforestation
apply	dilemmas

1) In many districts cattle were thought essential for rice cultivation, and when there was a shortage, fields __________ .

2) Without more oversight, farming for biodiesel could __________ worldwide.

3) Intelligent watering technologies have been developed in recent years to ___________ to turf and landscape plants.

4) The agricultural land in this region ___________ due to excessive irrigation and poor drainage, leading to reduced cro yields.

5) The new technology ___________ for policymakers, who must balance innovation with ethical concerns.

Section III Sharing Your Ideas

Task 8 ***The shortsighted demands of the Sumerian rulers led to the collapse of their civilization. What would you do if you were a ruler of Sumer? What measures would you recommend to avoid the negative impacts described in the text without endangering the economy of the country?***

Active Reading 2

Warming Up

Task ***What was life like for the ancient Egyptians? Listen to a song about ancient Egypt, and discuss the following questions with your partners.***

1) What were ancient Egyptian houses like?

2) What kinds of food did the ancient Egyptians eat?

3) Can you imagine what life would be like for the Egyptians without the Nile River?

Reading

Agriculture and Irrigation

1 Eleven thousand years ago, the inhabitants of the Nile Valley pursued hunting, gathering and fishing. By 5850 B.C., goats and sheep from the Levantine cradle were brought to the deserts surrounding the Nile Valley, and by 4795 B.C., communities of farmers, who were cultivating wheat and barley and were herding sheep, goats and cattle, began to emerge in the Nile Delta. Egypt was on its way to becoming one of the major agricultural centers of the ancient world. As Adolf Erman put it, "Agriculture is the foundation of Egyptian civilization". Ancient Egypt was an agrarian rather than an urban society.

2 Egyptian agricultural sustainability was provided by the deposition of fertile alluvial soil containing mineral traces of organic debris brought down in the flood from the mountains and swamps further south. The Greek historian Herodotus, observing that the soil of Egypt had been formed by the river's sediment, pronounced Egypt to be the "gift of the Nile". The Egyptians were aware of this: an early monument reads, "The Nile supplies all the people with nourishment and food". Their environment encouraged them to think of

processes of nature as operating in predictable cycles. The Nile flooded its banks at almost the same time every year (beginning in late July or early August). The only fertile land was what the river watered in the long, narrow valley floor of Upper Egypt and the broad Delta of Lower Egypt. For millennia, much of Egypt's food has been cultivated in the Nile delta region. The Egyptians grew a variety of crops for consumption, including grains, vegetables and fruits. However, their diets revolved around several staple crops, especially cereals and barley.

3 The flood was not totally predictable: a high Nile might wash away villages and spread desolation over the land, or a low Nile might fail to water the land adequately. In some periods when the river failed, rebels or invaders took advantage of weakness and unrest. As a result, Egyptian history was punctuated by times when pharaonic government collapsed. But traditional patterns of environmental relationships reappeared with phenomenal tenacity. As John Wilson expressed it, "The Nile never refused its great task of revivification. In its periodicity it promoted the [Egyptians'] sense of confidence; in its rebirth it gave [them] a faith that [they], too, would be victorious over death and go on into eternal life. True, the Nile might fall short of its full bounty for years of famine, but it never ceased altogether, and ultimately it always came back with full prodigality".

4 But the Egyptians also had a useful, sophisticated technology that kept their civilization operating well. That was the system of water management called basin irrigation that used the natural flooding of the Nile River, with irrigation works and careful planning, to keep the agricultural base functioning. It was an appropriate technology for the ecological situation of a rainless land watered by an exotic river flowing from East Africa.

5 Technological inventions were made, such as the *shaduf*, a bucket on a long-counterbalanced arm. Nilometers were installed near the First Cataract and elsewhere to measure the height of the river and to help predict the extent of the annual flood. Egypt incorporated such advances into the system of environmental regulation.

6 The Egyptians lacked science in the modern sense. They expressed an understanding of the workings of nature in religious images, and they explained technology in terms of the sacred. In this perspective, irrigation was an activity originated by the gods. Sacred geometry, sacred astronomy, and sacred records were marshaled to assure what we would call sustainability. Geometry, elaborated through trial and error to reestablish boundaries between fields when markers had been swept away in the flood, was regarded as a hallowed occupation devised by the wise god Thoth and entrusted to trained priest-scribes. Temples were oriented to keep watch on the revolutions of the sun and stars, which would tell when to open canals. Papyri containing these arcane branches of knowledge were kept in temple libraries.

7 Indeed, early pieces of art show that irrigation was practiced by the pharaoh himself. The first-dynasty Scorpion-King mace head shows the king digging a canal, and "Canal-digger" was an important administrative title. Canal building was believed to be a major occupation of those in the blessed world beyond death. Some scholars think that the monarchy of the pharaoh was an outgrowth of the need to direct hydro-engineering on a country-wide scale, although most irrigation work was supervised by local officials in the names, districts the size of American counties.

8 Irrigation works extended cropland area beyond the area naturally flooded. The two types of land were kept distinct: *Rei* fields were those ordinarily covered by flood; *Sharaki* land required artificial irrigation. Laborers dredged channels, dug ditches, built dams, constructed dykes and levees, and used buckets to raise water. These activities were considered parts of a holy occupation. Major projects sponsored by pharaohs were commemorated as good works; Pepi Ⅰ (2390-2360 BC), for instance, cut a canal to water a new district.

9 Continual watchfulness and constant labor and care were needed in order to maintain these works. Otherwise the canals would quickly silt up and become useless or the levees would be washed away by the pressure of the water. Since Egypt was primarily an agricultural country, the measure in which this authority was maintained was the measure of the prosperity of the nation. Every lapse in governmental efficiency thus bespoke a corresponding diminution of economic wealth. In times of breaking dynasties when there was no strong authority the canals fell into disuse, and hard times for the people followed. A reliable food supply allowed overpopulation. When population increased to near the highest level that could be supported in a year of good harvest, any abnormally low harvest would bring the danger of famine. Reliefs on the causeway of Unas at Sakkara show people starving, their ribs conspicuous. The Twenty-second Dynasty was one of declining fortunes and from it has come the following account: "The flood came on in this whole land; it invaded the two shores as in the beginning. This land was in his power like the sea, there was no dike of the people to withstand its fury.

(Adapted from Hughes, 2009; Hassan, 2005; Olmstead, 1917)

New Words and Expressions

herd /hɜːd/ *vt.* 驱赶(兽群),放牧

agrarian /əˈgreərɪən/ *adj.* 土地的

deposition /ˌdepəˈzɪʃn/ *n.* 沉积,沉淀(物)

alluvial /əˈluːvɪəl/ *adj.* 冲积的

debris /ˈdeɪbriː/ *n.* 泥石,岩屑

swamp /swɔmp/ *n.* 沼泽,湿地

sediment /ˈsedɪmənt/ *n.* 沉淀物,沉渣

nourishment /ˈnʌrɪʃmənt/ *n.* 滋养品,营养

punctuate /ˈpʌŋktʃueɪt/ *vt.* 打断

pharaonic /ˌfeərɪˈɔnɪk/ *adj.* 古埃及法老王的

tenacity /təˈnæsətɪ/ *n.* 顽强,执着,韧劲

famine /ˈfæmɪn/ *n.* 饥荒

cease /siːs/ *vi.* 终止，结束

prodigality /ˌprɔdɪˈɡælətɪ/ *n.* 丰富，慷慨

exotic /ɪɡˈzɔtɪk/ *adj.* 外来的，异国的，不一样的

install /ɪnˈstɔːl/ *vt.* 安装，设置

sacred /ˈseɪkrɪd/ *adj.* 神圣的，与神有关的

marshal /ˈmɑːʃ(ə)l/ *vt.* 整理(思想，思路等)

hallowed /ˈhæləʊd/ *adj.* 神圣的，神圣化的

entrust /ɪnˈtrʌst/ *vt.* 委托，交托

commemorate /kəˈmeməreɪt/ *vt.* 用以纪念

levee /ˈlevɪ/ *n.* 堤坝

lapse /læps/ *n.* (两件事发生的)间隔时间

conspicuous /kənˈspɪkjuəs/ *adj.* 明显的，显眼的

Exercises

Section I Knowledge Focus

Task 1 ***Read Paras. 1-6 and complete the following diagram by filling in the blanks with words from these paragraphs.*** (***ONE WORD ONLY***)

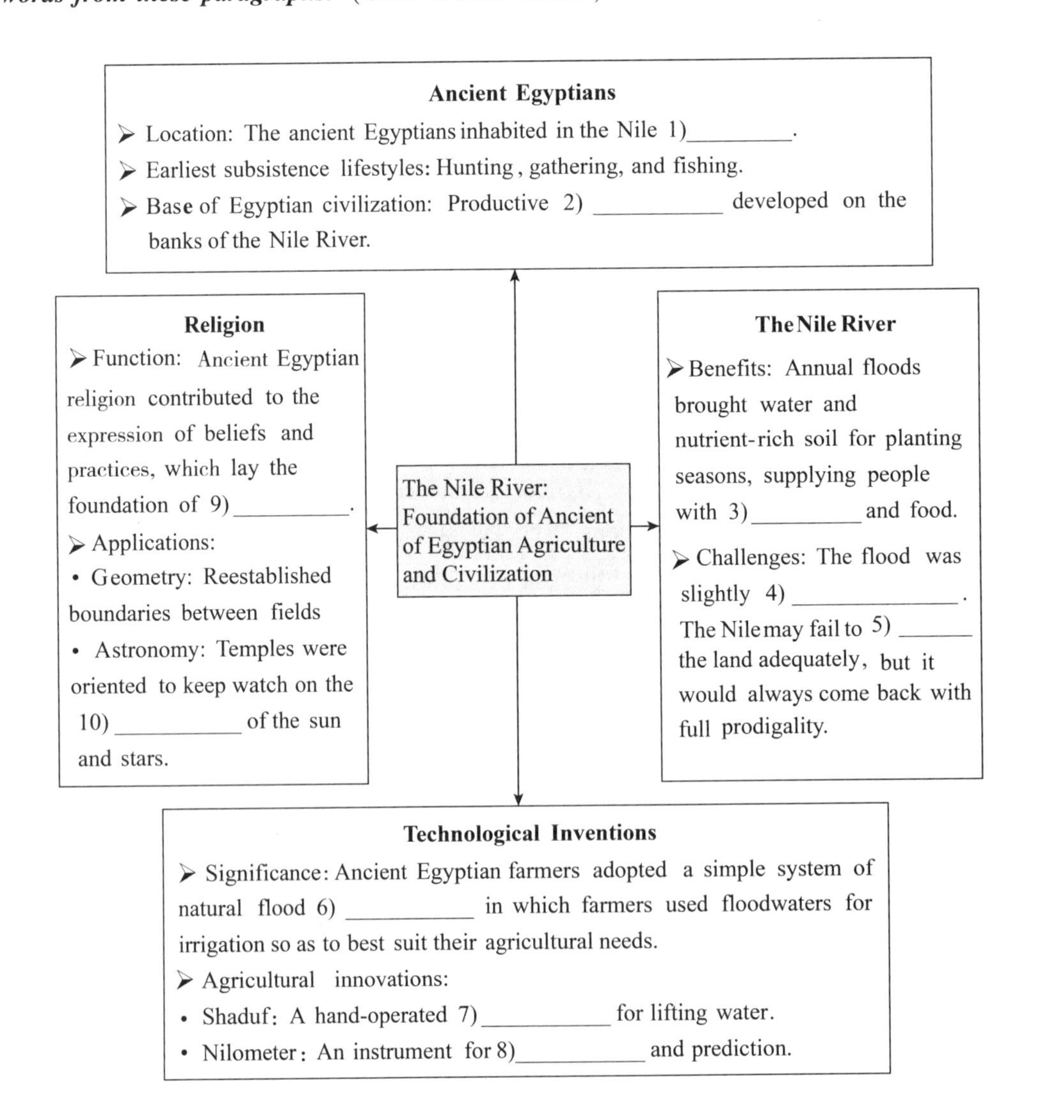

Task 2 ***Read Paras. 7-9 and find out FIVE false descriptions in the following diagram, and correct them.***

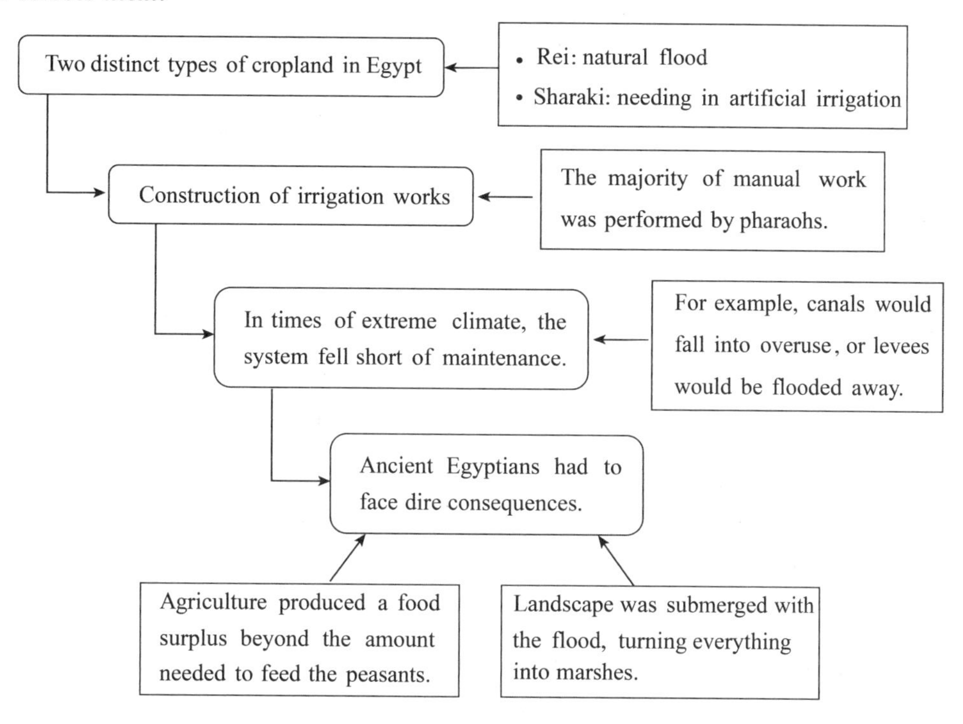

Task 3 ***Given below are five statements. Each statement contains information given in one of the paragraphs of the text. Identify the paragraph from which the information is derived. Answer the questions by writing down the paragraph number (1-9) for each statement.***

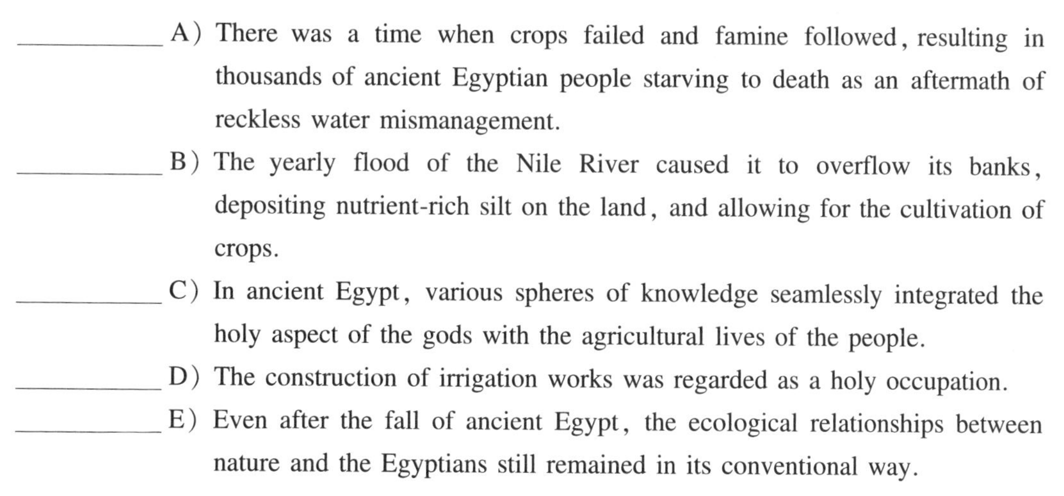

_____ A) There was a time when crops failed and famine followed, resulting in thousands of ancient Egyptian people starving to death as an aftermath of reckless water mismanagement.

_____ B) The yearly flood of the Nile River caused it to overflow its banks, depositing nutrient-rich silt on the land, and allowing for the cultivation of crops.

_____ C) In ancient Egypt, various spheres of knowledge seamlessly integrated the holy aspect of the gods with the agricultural lives of the people.

_____ D) The construction of irrigation works was regarded as a holy occupation.

_____ E) Even after the fall of ancient Egypt, the ecological relationships between nature and the Egyptians still remained in its conventional way.

Section II Language Focus

视听资源

Task 4 ***Watch a video clip about Egyptian farming and complete the following sentences below based on what you hear.*** (***NO MORE THAN THREE WORDS***)

Ancient Egyptians 1) ________ all nutrients left behind by the 2) ________. The extent of the flood can also be 3) ________ by measuring the level of the river. During the growing season, the fields are 4) ________ either manually or with oxen. After harvesting, the food is held in big stores to keep it 5) ________.

Task 5 ***Identify what is being described in the following phrases. Choose your answer from the box below and write the corresponding word in the space provided.***

insulation	innovation	famine	unrest	outgrowth

________ 1) extreme shortage of food in a region.

________ 2) detachment from other objects.

________ 3) a state of trouble and confusion, especially in a political sense.

________ 4) the introduction of new ideas or methods.

________ 5) a natural development or result of something.

Task 6 ***Complete the following sentences with appropriate words or expressions given below. Change the form where necessary.***

orient	exotic	tenacity	sacred	devise

1) The environmental activist's ________ efforts to promote sustainable practices have significantly influenced public policy, leading to a remarkable reduction in the city's carbon footprint.

2) It was the ancient Egyptian gods and priests who ________ the holy practice of: geometry.

3) Cats are perhaps the most ________ of all Ancient Egyptian animals.

4) The majority of ancient Egyptian works were never ________ to be seen—but to benefit a divine or deceased recipient.

5) In the dream she eats ________ and runs to swim in the ocean.

cease	dredge	install	sponsor	silt

6) When the Old Egypt Kingdom broke down, the maintenance of agricultural systems also ________.

7) The company has been offering to ________ children from under-privileged regions at university.

8) The old harbor was ________ up centuries ago.

9) Fearing more floods, the Kingdom had the river ________.

10) Dickens wrote his novels in weekly ________ for a magazine.

Task 7 ***Translate the following Chinese expressions into English with what you have learned from the text.***

1) 排水系统　2) 洪水多发区　3) 有机农业　4) 生态破坏　5) 农作物多样性
6) 可持续农业　7) 水危机　8) 土壤侵蚀　9) 防止沙漠化　10) 土壤退化

Section III Sharing Your Ideas

Task 8 ***Compare the two earliest agrarian civilizations discussed in Active Reading 1 and 2. Complete the chart below to explore their diversity and environmental interactions.***

Aspects of Civilization	**Sumer**	**Ancient Egypt**
Geographical features		
Agricultural patterns (e. g. main crops)		
Significance of agriculture		
Environmental challenges and impacts		
Environmental alteration due to agricultural practices		

Part Three　China's Environmental Story

Active Reading 3

Warming Up

Task ***Discuss the following questions with your partners.***

1) Bamboo cages filled with cobblestones were used in the Dujiangyan Irrigation System. Why do you think ancient engineers chose bamboo and cobblestones over other materials?

2) If you were to design a similar water management project today, what everyday materials would you choose?

Reading

Dujiangyan Irrigation System: Ancient Chinese Flood Management Wisdom

1 Every year in August, heavy rainstorms hit Leshan, Sichuan Province, and the Minjiang River's water level at the Leshan Section always rises above the warning line this time of year.

2 It isn't just Leshan City; many parts of Sichuan Province suffer from floods, but Dujiangyan, also located along the Minjiang River, remained unscathed. The ancient irrigation system constructed about 2,000 years ago still works today.

3 Located on the Minjiang River in Sichuan Province, west of the Chengdu Plain, the Dujiangyan Irrigation System (DIS) is one of the oldest irrigation systems in the world and it is also the only surviving monumental non-dam irrigation system from ancient times. With a history of over 2,000 years, its continuous function has turned it into a living relic, benefitting people living on the Chengdu Plain and making many believe that it is actually a greater architectural miracle than the Great Wall. This system is still used to irrigate over 668,700 hectares of farmland, drain floodwater, and it provides water to more than 50 cities in Sichuan province today.

Original Construction

4 Why was this huge project built?

5 There are three main reasons for Dujiangyan. First of all is flood control, second is to provide a stable strategic rear base for the national reunification, and also for the development of water transport routes.

6 Li Bing, governor of Shu for the state of Qin, and his son led the construction. Instead of building a dam, they harnessed the river using a novel method for that time: dividing the water.

7 There are three main parts of the construction: the Yuzui (Fish Mouth Levee), the Feishayan (Flying Sand Weir) and the Baopingkou (Bottle-Neck Channel). Each has a different, indispensable function.

8 The first Minjiang River barrier is Yuzui. It was built at the bend of the river, where the surging water is divided into the inner and outer rivers by the "dikes". The lateral canal drains the flood and the inner canal flows into the Chengdu Plain through Baopingkou.

9 The water that runs through Yuzui is less turbulent, but it still carries a lot of sand, and that's when Feishayan works. The weir has an opening that connects the inner and outer streams which allows the swirling flow to drain out excess water. Every year, the river workers clean the sand up, which effectively prevents the silting of the river from breaking its banks.

10 The place where Baopingkou is located was Yulei Mountain. It was named Baopingkou (Bottle-Neck Channel) because of its shape. It took Li Bing and his team eight years to chisel the stone wall open to form the constant width for a thousand years. The purpose of Baopingkou was to divert and irrigate the water entering the Chengdu plain. If a large amount of water was blocked by the Baopingkou during the flood period, the water level would rise. When the water level exceeded a certain level, the drainage channel behind the Feishayan weir would be discharged to the outer river to achieve the secondary flood discharge.

11 Every design in the Dujiangyan irrigation system makes full use of the local environmental characteristics. It takes advantage of nature instead of conquering it by means of human intervention with minimal impact.

Maintenance & Development

12 Through trial and error, war and natural disasters, the irrigation system project gradually evolved to adapt to the changing environment.

13 Unfortunately, the Dujiangyan Irrigation System has been wrecked several times due to war and natural disasters. Surprisingly, however, although the irrigation system has been wrecked on multiple occasions it has found its way to recover quickly because of its simplistic architecture and design.

14 Originally, the levees of the Dujiangyan Irrigation System were built with bamboo cages filled with cobblestones. The simplicity of the design and materials enabled engineers to replace or remove issue spots with ease. When the DIS was destroyed the natural rustic

build of the irrigation's structure allowed engineers to make full repairs within the same year it was destroyed. In addition to the simple architecture, the tools used to create the levee were also very basic. To create the system all that was used was a hoe and a knife. The unsophistication of these traditional materials is the secret to the efficiency and convenience of the DIS. The tools and materials used in the creation of this irrigation system are abundant and can easily be found in any local area, even in under-developed agricultural societies. Moreover, unlike other water systems, because of the simplicity of the tools, no pollution was released into the air during this process. This practice of the use of simple materials and tools allowed high utilization and efficiency of the materials.

Historical Significance

15 The success of this water conservancy project is due to the comprehensive and systematic strategy and thinking. The consistency of harmony between man and nature is what has allowed the Dujiangyan System to continuously be mended and changed. Since the establishment of the Dujiangyan Irrigation System about 2200 years ago, DIS has been a key infrastructure in rice cultivation.

16 The Dujiangyan Irrigation System has a rich history of evolution and modification, allowing it to endure for centuries. It has been instrumental in making Chengdu an economic center and has brought water to areas that previously lacked a source of water, thus creating fields of rich cultivation where there wasn't before. Unlike other dam projects, the DIS did not displace the villages living around the dam site.

17 Although this is one of the oldest and best designed systems, many people around the world have no knowledge of the Dujiangyan Irrigation System. More citizens globally should be aware of the history and amazing benefits this irrigation system has provided the Chinese for so many decades. In 2000, the Dujiangyan Irrigation System was deemed a World Heritage Site by the United Nations Educational, Scientific, and Cultural Organization (UNESCO). Scientists around the world admire this irrigation project because Dujiangyan allows water to flow over the dam naturally, enabling humans and nature to co-exist peacefully.

(Adapted from Price et al.,2019;Meng et al.,2020)

New Words and Expressions

unscathed /ʌnˈskeɪðd/ *adj.* 未受伤的

rear /rɪə/ *adj.* 后部的

reunification /ˌriːjuːnɪfɪˈkeɪʃn/ *n.* 重新统一

harness /ˈhɑːnɪs/ *vt.* 控制并利用

indispensable /ˌɪndɪˈspensəbl/ *adj.* 不可或缺的,必须的

bend /bend/ *n.* (道路或河流的)拐弯,弯道

dike /daɪk/ *n.* 堤坝,大坝

turbulent /ˈtɜːbjulənt/ *adj.* (水)湍急的

weir /wɪə/ *n.* 堰,坝

swirling /ˈswɜːlɪŋ/ *adj.* 打旋的
exceed /ɪkˈsiːd/ *vt.* 超过,超出
wreck /rek/ *vt.* (严重)破坏,摧毁(建筑)
cobblestone /ˈkɔblstəʊn/ *n.* (铺设街道用的)圆形鹅卵石
rustic /ˈrʌstɪk/ *adj.* 乡村的;纯朴的
exert /ɪgˈzɜːt/ *vt.* 运用,施加(影响)
conservancy /kənˈsɜːvənsɪ/ *n.* 管理;保护;保存
mend /mend/ *vt.* 修理,修补

Exercises

Section I Understanding the Text

Task 1 ***Discuss the following questions in small groups.***

1) Why, among all the cities along the Minjiang River, was Dujiangyan the only one that did not suffer from heavy storms and floods?
2) In terms of its structure, what are the main features of the Dujiangyan Irrigation System?
3) What makes the Dujiangyan Irrigation System so easy to maintain?
4) What is the main philosophy behind the construction of the Dujiangyan Irrigation System?

Section II Developing Critical Thinking

Task 2 ***Telling China's stories effectively can help shape a true and comprehensive image of China, countering misinformation and enhancing international understanding. In this context, there has been a strong emphasis on sharing China's values, focusing on its truths, virtues and aesthetic achievements with the global community.***

Imagine that you are at an international seminar, invited to give a presentation on Chinese wisdom regarding human-nature coexistence, one of the fundamental principles in the development of China's civilization. You may prepare your presentation by following the instructions below.

Step 1 Discover your topic

Gather as much information as possible about your topic.

Step 2 Structure your presentation

Craft your message in a logical and simple way. Below is a natural flow that any well-structured presentation should follow.

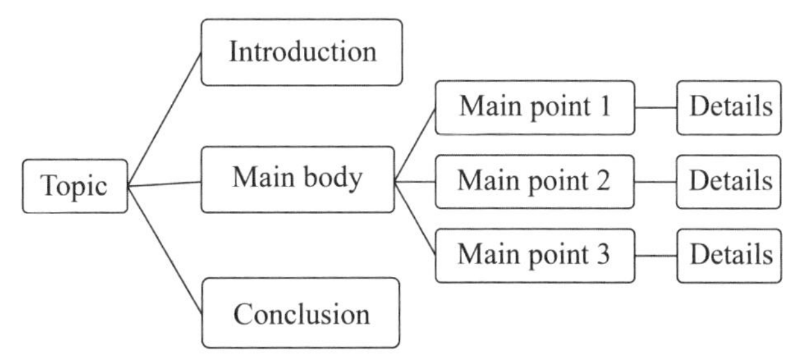

Step 3 Deliver your presentation

Now that you have everything prepared and organized, you are ready to present. You may refer to the following checklist to make your presentation effective and impressive.

1) The topic is introduced clearly in the opening.
2) The structure of the presentation is well-organized.
3) No essential information is missing.
4) There are details to support each main point.
5) Words and expressions are used appropriately. Try to use the ones you have learned in this unit.
6) Find the pronunciation of the words unfamiliar to you in a dictionary.
7) Have eye contact with the audience and use appropriate gestures.
8) Pay attention to the voice volume, tone and pacing to make yourself easy to follow.

UNIT 4

Changes in Biosphere in Early Modern Period

Part One Preparation

Unit Preview

Throughout history, human activities have played a crucial role in shaping the natural world. With the rise of global trade and cultural exchange, new species and practices have been introduced to different parts of the world, significantly altering the environment.

The Columbian Exchange, a transformative period initiated by Christopher Columbus' voyages in the late 15th and early 16th centuries, led to the widespread transfer of plants, animals, and diseases between the Old and New Worlds. While it enriched global diets and cultures, it also caused devastating disease outbreaks and ecological disruptions, with impacts that persist today.

The introduction of European rabbits to Australia in the 19th century, imported for sport, triggered an ecological catastrophe, decimating native flora and fauna. Despite extensive efforts to control their population, their destructive impact continues to shape Australia's environment.

The history of the soybean, from its origins in ancient China to its global cultivation, reveals its transformation from a humble food for the poor to a crucial agricultural commodity with significant economic and cultural influence worldwide.

These case studies underscore the complex interplay between human actions and the natural world, highlighting the critical need to understand the long-term effects of introducing species and practices to new environments, offering valuable lessons for sustainable management in an interconnected world.

Learning Objectives

Upon completion of this unit, you will be able to:

- develop a deeper understanding of the changes in biosphere during the early modern period.
- evaluate the environmental and societal impacts of the Columbian Exchange.
- analyze the relationship between changes in biosphere and the human exchange activities.

Part Two Global Perspectives

Active Reading 1

Warming Up

Task 1 ***Answer following questions.***

1) Who discovered the New World?

2) What else do you know about the discoverer?

Task 2 ***Watch a short video clip. Then discuss the following questions with your partners.***

1) Why was Columbus fascinated by the idea of sailing west from Europe to Asia?

2) When did Columbus start his first voyage?

3) What is the impact of the voyages of Columbus?

Reading

The Columbian Exchange

1 La Isabela is situated on the north side of the great Caribbean island of Hispaniola, in what is now the Dominican Republic. It was the initial attempt by Europeans to make a permanent base in the Americas, where locates the first American residence of Christopher Columbus, the man whom generations of schoolchildren have learned to call the discoverer of the New World. (To be precise, La Isabela marked the beginning of consequential European settlement—Vikings had established a short-lived village in Newfoundland five centuries before) By founding La Isabela, Columbus initiated permanent European occupation in the Americas. And in so doing he began the era of globalization—the single, turbulent exchange of goods and services that today engulfs the entire habitable world.

2 Long-distance trade had occurred for more than a thousand years before Columbus' voyage, much of it across the Indian Ocean. China had for centuries sent silk to the Mediterranean by the Silk Road, a route that was lengthy, dangerous, and, for those who survived, hugely profitable. But nothing like this worldwide exchange started by Columbus had existed before, still less sprung up so quickly, or functioned so continuously. No previous trade networks included both of the globe's two hemispheres; nor had they operated on a scale large enough to disrupt societies on opposite sides of the planet.

3 Newspapers usually describe globalization in purely economic terms, but it is also a biological phenomenon; indeed, from a long-term perspective it may be primarily a

biological phenomenon. After 1492 the world's ecosystems collided and mixed as European vessels carried thousands of species to new homes across the oceans. The Columbian Exchange, as Crosby called it, is the reason there are tomatoes in Italy, oranges in the United States, chocolates in Switzerland, and chili peppers in Thailand. To ecologists, the Columbian Exchange is arguably the most important event since the death of the dinosaurs.

4 Despite the brevity of its existence, La Isabela marked the beginning of an enormous change: the creation of the modern Caribbean landscape. Columbus and his crew did not voyage alone. They were accompanied by a menagerie of insects, plants, mammals, and microorganisms. Beginning with La Isabela, European expeditions brought cattle, sheep, and horses, along with crops like sugarcane (originally from New Guinea), wheat (from the Middle East), bananas (from Africa), and coffee (also from Africa). Equally important, creatures the colonists knew nothing about hitchhiked along for the ride. Earthworms, mosquitoes, and cockroaches; honeybees, dandelions, and African grasses; rats of every description—all of them poured from the hulls of Colón's vessels and those that followed, rushing like eager tourists into lands that had never seen their like before.

5 Cattle and sheep ground American vegetation between their flat teeth, preventing the regrowth of native shrubs and trees. Beneath their hooves would sprout grasses from Africa, possibly introduced from slave-ship bedding; splay-leaved and dense on the ground, they choked out native vegetation. (Alien grasses could withstand grazing better than Caribbean groundcover plants because grasses grow from the base of the leaf, unlike most other species, which grow from the tip. Grazing consumes the growth zones of the latter but has little impact on those in the former) Over the years forests of Caribbean palm, mahogany, and ceiba became forests of Australian acacia, Ethiopian shrubs, and Central American logwood. Scurrying below, mongooses from India eagerly drove Dominican snakes toward extinction. The change continues to this day. Orange groves, introduced to Hispaniola from Spain, have recently begun to fall to the depredations of lime swallowtail butterflies, citrus pests from Southeast Asia that probably came over in 2004. Today Hispaniola has only small fragments of its original forest.

6 In the New World, natives and newcomers interacted in unexpected ways, creating biological bedlam. When Spanish colonists imported African plantains in 1516, the Harvard entomologist Edward O. Wilson has proposed, they also imported scale insects, small creatures with tough, waxy coats that suck the juices from plant roots and stems. About a dozen banana-infesting scale insects are known in Africa. In Hispaniola, Wilson argued, these insects had no natural enemies. In consequence, their numbers must have exploded—a phenomenon known to science as "ecological release". The spread of scale insects would have dismayed the island's European banana farmers but delighted one of its

native species: the tropical fire ant *Solenopsis geminata*. *S. geminata* is fond of dining on scale insects' sugary excrement; to ensure the flow, the ants will attack anything that disturbs them. A big increase in scale insects would have led to a big increase in fire ants.

7 In 1518 and 1519, according to Bartolomé de Las Casas, a missionary priest who lived through the incident, Spanish orange, pomegranate, and cassia plantations were destroyed "from the root up". Thousands of acres of orchards were "all scorched and dried out, as though flames had fallen from the sky and burned them". The actual culprit, Wilson argued, was the sap-sucking scale insects. But what the Spaniards saw was *S. geminata*— "an infinite number of ants", Las Casas reported, their stings causing "greater pains than wasps that bite and hurt men". The hordes of ants swarmed through houses, blackening roofs "as if they had been sprayed with charcoal dust", covering floors in such numbers that colonists could sleep only by placing the legs of their beds in bowls of water. They "could not be stopped in any way nor by any human means".

8 From the human perspective, the most dramatic impact of the Columbian Exchange was on humankind itself. European contact enabled the transmission of diseases to previously isolated communities, which caused devastation far exceeding that of even the Black Death in fourteenth-century. Europeans brought deadly viruses and bacteria, such as smallpox, measles, typhus, and cholera, for which Native Americans had no immunity. Spanish accounts suggest that Hispaniola had had a large native population: Columbus, for instance, casually described the Taino as "innumerable, for I believe there to be millions upon millions of them". Las Casas claimed the population to be "more than three million". Modern researchers have not nailed down the number; estimates range from 60,000 to almost 8 million. A careful study in 2003 argued that the true figure was "a few hundred thousand". No matter what the original number, though, the European impact was horrific. In 1514, twenty-two years after Columbus' first voyage, the Spanish government counted up the Indians on Hispaniola for the purpose of allocating them among colonists as laborers. Census agents fanned across the island but found only 26,000 Taino. Thirty-four years later, according to one scholarly Spanish resident, fewer than 500 Taino were alive.

9 Spanish cruelty played its part in the calamity, but its larger cause was the Columbian Exchange. Before Columbus, none of the epidemic diseases common in Europe and Asia existed in the Americas. Throughout the sixteenth and seventeenth centuries novel microorganisms spread across the Americas, ricocheting from victim to victim, killing three-quarters or more of the people in the hemisphere. It was as if the suffering these diseases had caused in Eurasia over the past millennia were concentrated into a span of decades. In the annals of human history there is no comparable demographic catastrophe. The Taino were removed from the face of the earth, though recent research hints that their

DNA may survive, invisibly, in Dominicans who have African or European features, genetic strands from different continents entangled, coded legacies of the Columbian Exchange.

(Adapted from Mann, 2012)

New Words and Expressions

engulf /ɪnˈgʌlf/ *vt.* 包围
habitable /ˈhæbɪtəb(ə)l/ *adj.* 适合居住的
hemisphere /ˈhemɪsfɪə(r)/ *n.* (地球的)半球
vessel /ˈves(ə)l/ *n.* 大船,轮船
menagerie /məˈnædʒərɪ/ *n.* (一群)野生动物
microorganism /ˌmaɪkrəʊˈɔːgənɪzəm/ *n.* 微生物
expedition /ˌekspəˈdɪʃ(ə)n/ *n.* 考察队
hull /hʌl/ *n.* 船身,船体
splay /spleɪ/ *n.* 展开;斜面
scurry /ˈskʌrɪ/ *vi.* 碎步疾跑
mongoose /ˈmɔŋguːs/ *n.* 獴(居于热带地区,捕食蛇、鼠等)
grove /grəʊv/ *n.* 果园
depredation /ˌdeprəˈdeɪʃn/ *n.* 破坏
bedlam /ˈbedləm/ *n.* 混乱嘈杂的场面
plantain /ˈplæntɪn/ *n.* 大蕉
entomologist /ˌentəˈmɔlədʒɪst/ *n.* 昆虫学家
sugary /ˈʃʊgərɪ/ *adj.* 含糖的,甜的
excrement /ˈekskrɪmənt/ *n.* 粪便,排泄物
missionary /ˈmɪʃən(ə)rɪ/ *adj.* 传教的
orchard /ˈɔːtʃəd/ *n.* 果园,果树林
scorch /skɔːtʃ/ *vt.* 使枯萎
culprit /ˈkʌlprɪt/ *n.* 肇事者;引起问题的事物
sap /sæp/ *n.* (植物体内运送养分的)液,汁
horde /hɔːd/ *n.* 一大群
swarm /swɔːm/ *v.* 成群地飞来飞去
exceed /ɪkˈsiːd/ *vt.* 超过(数量)
innumerable /ɪˈnjuːmərəb(ə)l/ *adj.* 无数的
census /ˈsensəs/ *n.* 人口普查
calamity /kəˈlæmətɪ/ *n.* 灾难,灾祸
epidemic /ˌepɪˈdemɪk/ *adj.* 流行的
ricochet /ˈrɪkəʃeɪ/ *vi.* (运动的物体)弹开;反弹出去
annal /ˈæn(ə)l/ *n.* 年鉴
demographic /ˌdeməˈgræfɪk/ *adj.* 人口的,人口统计的
entangle /ɪnˈtæŋgl/ *v.* 使纠缠,缠住;套住
legacy /ˈlegəsɪ/ *n.* 遗留问题;后遗症

Exercises

Section I Knowledge Focus

Task 1 ***Read Paras. 1-2 and complete the following diagram by filling in the blanks with words from these paragraphs. (ONE WORD ONLY)***

Timeline	People / Country	Exchange Activities
1,000 years before Columbus' voyage	China	...much of it across the Indian Ocean. China had for centuries sent silk to the Mediterranean by the 1)__________ Road.
500 years before Columbus' voyage	2) __________	They had established a short-lived 3) __________ in Newfoundland.
4) __________-1504	Christopher Columbus	By founding La Isabela, Columbus initiated permanent European 5) __________ in the Americas.

Task 2 ***Read Paras. 4-9 and complete the following diagram illuistrating the Columbian Exchange by filling in the blanks with words from these paragraphs. Change the form where necessary. (ONE WORD ONLY)***

Plants

➢ Cause: European settlers brought crops from Europe, Africa, and Asia, e.g., sugarcane, wheat, bananas, etc.

➢ Effects:

- Alien grasses from Africa sprouted and grew densely, 1)__________out native vegetation
- New crops out-competednative plants, and today Hispaniola has only small fragments of its 2)________forest.

Diseases

➢ Cause: Europeans brought deadly 6)_______ and bacteria, such as smallpox, measles, typhus, and cholera, for which Native Americans had no 7)_______.

➢ Effects:

- The Taino, whose population had once been 8)_____, were 9)____ from the face of the earth.
- There is no comparable 10)_____ catastrophe in human history.

Animals

➢ Cause: Along with plants, European settlers brought animals, e.g., cattle, mongoose, earthworms, etc.

➢ Effects:

- Cattle and sheep grazing prevented the 3)________ of native shrubs and trees.
- Mongooses from India eagerly drove Dominican snakes toward 4)________.
- Crop destruction in 1518-1519 resulted from the ecological 5)________ effect of scale insects.

Task 3 ***Based on your understanding of the passage, decide whether the following statements are true or false. Put T for true and F for false in the blank provided before each statement.***

________ 1) For more than a thousand years before Columbus' voyage, people who survived the Silk Road, a long and dangerous route, made a great fortune by sending silk to the Mediterranean.

________ 2) From a long-term perspective, globalization is more of an economic term than a biological phenomenon.

________ 3) Nowadays, the original forests in Hispaniola still flourish because grazing has had little impact on them.

________ 4) From a human perspective, the most dramatic impact of the Columbian Exchange was on animals and plants.

________ 5) Throughout the sixteenth and seventeenth centuries, epidemic diseases killed three-quarters or more of the people in the Americas, which is considered an incomparable demographic catastrophe in the annals of human history.

Section II Language Focus

Task 4 ***Match the term in the left column with an explanation given in the right column and write the corresponding letter in the space provided below.***

1) calamity	A. *n.* one half of the earth, especially the half above or below the equator
2) vessel	B. *n.* a collection of wild animals
3) hemisphere	C. *n.* a large ship or boat
4) menagerie	D. *n.* an event that causes great damage to people's lives, property, etc.

1) ________ 2) ________ 3) ________ 4) ________

Task 5 ***Complete the following sentences with appropriate words given below. Change the form where necessary.***

calamity legacy swarm turbulent consequential

1) He who guards his mouth and his tongue keeps himself from ________.

2) People ________ to the stores, buying up everything in sight.

3) The aircraft is designed to withstand ________ conditions.

4) Early childhood experiences can be very ________ for children's long-term social, emotional and cognitive development.

5) Future generations will be left with a ________ of pollution.

culprit initiate engulf ricochet depredation

6) Much of the region's environmental ________ is a result of poor planning.

7) People who are talked to have equally positive experiences as those who ________ a conversation.

8) Someday the euro debt crisis that started in Greece and spread to ________ Europe will be over.

9) The bullet ________ off a nearby wall.

10) The main ________ in the current crisis seems to be modern farming techniques.

Task 6 ***Find the words in the box that have the same meaning as the underlined words or phrases in the following sentences, and write the corresponding word in the space provided.***

innumerable exceed expedition sugary scorch

________ 1) If you go beyond your credit limit, we have the right to suspend or cancel your account.

________ 2) Rapid urbanization brings with it a more westernized and generally more sweet diet.

________ 3) This is just one example of the countless exploits of this type.

________ 4) The leaves will wither if you water them in the sun.

________ 5) I knew the rest of the excursion would be on foot, and I'd hope to be leaving the bike at this place.

Task 7 ***Match the words in the left column with those in the right column to form appropriate expressions. Then complete the following sentences with one of the expressions. Change the form or add articles where necessary.***

scale	release
demographic	disease
ecological	catastrophe
epidemic	insect

1) The Japanese firms face a ________, and the solution is to treat women better.

2) The strategy may not be so good for the few males left in ________ populations: They're becoming obsolete and may eventually go extinct.

3) The outbreak of COVID-19 has become a global ________.

4) Dr. Olsen and his colleagues suggest that the explanation for this rapid increase in size may be a phenomenon called ________.

Section III Sharing Your Ideas

The Columbian Exchange has introduced many new species, crops, and animals to different parts of the world, but also caused significant environmental and social disruptions.

Task 8 ***Can you think of other examples where human actions led to unintended ecological consequences? What lessons can we learn from history?***

Active Reading 2

Warming Up

Task ***From 1902 to 1907, Australian government launched the third battle against rabbits by building the longest fence in the world. Discuss the following questions with your partners.***

1) Can you guess the result of the battle according to the living habits of the rabbits?

2) Why have rabbits been a disaster in Australia according to your knowledge?

Reading

The Great Australian Rabbit Disaster

1 The feral European rabbit is one of the most widely distributed and abundant mammals in Australia. It causes severe damage to the natural environment and to agriculture.

2 In the 19th century, settlers from Europe came in droves to this "new wonderful land", Australia. The settlers brought with them many customs from home. One of those customs was rabbit hunting; but unfortunately, Australia did not have any native rabbit species. So in 1859, a European landowner named Thomas Austin imported 24 wild rabbits from England and released them onto his land outside the town of Winchelsea, Victoria. With few rabbit predators and diseases, the rabbits reproduced prolifically. Within only 30 years, the original 24 rabbits had reproduced exponentially to millions and inhabited the states of Victoria, South Wales, and portions of Queensland and South Australia. Large areas of once relatively fertile landscapes were soon transformed into dry areas that became increasingly prone to topsoil loss and drought. In subsequent decades and throughout the majority of the 20th Century, rabbits have migrated to all corners of Australia and now inhabit even the most unlikely areas.

3 Feral rabbits can be found in many different habitats across Australia, ranging from deserts to coastal plains—wherever there is suitable soil for digging warrens. These rabbits are extremely adaptive, which has played a role in their spread across the Australian continent. All the rabbits need is soil that is fit to burrow and short grasses to graze on. Since these conditions are fairly easy to come by, they can adapt to new habitats such as the deserts and plains of Australia as easily as the meadows of Europe.

4 Not only are European rabbits adaptable creatures, they are also known for rapidly producing large quantities of offspring. Feral rabbits can breed from the age of four

months, and can do so at any time of the year, particularly when food is in good supply. In favorable conditions, they can produce five or more litters in a year, with four or five young in each litter. Even in unfavorable conditions, they can produce one or two litters a year.

5 Feral rabbits compete with native wildlife, damage vegetation and degrade the land. They ringbark trees and shrubs, and prevent regeneration by eating seeds and seedlings. Their impact often increases during drought and immediately after a fire, when food is scarce and they eat whatever they can. By removing above-ground and below-ground vegetation through grazing and warren construction, rabbits contribute to erosion—the loss of topsoil by wind and rain. In the Norfolk Island group, feral rabbits and goats reduced Philip Island to bedrock, leaving at least two plants locally extinct. "Rabbits are very good at finding the seedlings of shrubs when they are very small and grazing them out to the extent where the native shrubs are completely unable to regenerate", said Greg Mutze, a research officer at the Department of Water, Land and Biodiversity Conservation in South Australia, to the Australian Broadcasting Corporation.

6 The ecology of the witchetty bush (*Acacia kempeana*) provides an example of complex problems caused by rabbits. Witchetty bush is a shrub or tree found in arid areas of Australia, especially WA, SA and the NT, but also in Qld and NSW. It grows up to 5 metres, but is slow growing given the low and erratic rainfall throughout its distribution. Large Cossid moths lay eggs under the trees and the larvae feed on the sap of the Acacia roots. The plants also support birds (nesting and foraging), insects (pollinators), and ants (seed harvesters)—as well as any bilbies dining on the witchetty grubs. There is very little regeneration of witchetty bushes in areas where rabbits co-exist, and it can take 10-20 years before they are tall enough to be safe from destruction by rabbits. In the presence of rabbits the incidence of the trees declines and the average age rises. Several decades of effective rabbit control would be required for regenerating plants to be safe from renewed rabbit grazing pressure.

7 Feral rabbits may have caused the extinction of several small ground-dwelling mammals of Australia's arid lands, and have contributed to the decline in numbers of many native animals. Native animals, such as the pig-footed bandicoot and the greater bilby, have seen their numbers decline dramatically as well. Why? They're going after the same food sources as the rabbits, and just can't compete with an all-consuming bunny rabbit horde.

8 Since the introduction of rabbits to Australia, farm livestock populations have also been influenced. It was estimated in 1936 that the extermination of the rabbit population in New South Wales would liberate enough land to accommodate twelve to twenty million more sheep.

9 In addition to effects on livestock, adverse effects were also placed upon other traditional prey species. Soon after the introduction of rabbits, Australian authorities realized they had a problem, so they introduced European foxes into the environment. The authorities thought that since foxes naturally prey on rabbits, they would be the perfect solution to the rabbit problem. However, the foxes seemed to take advantage of their relatively exotic surroundings and turned their attention to more easily caught prey including rare ground birds, indigenous marsupials, and rodents. The only remaining semi-natural enemy of the rabbit population in Australia is now the dingo (a free-roaming wild dog). However, in the early 20th century, dingoes were frequently slaughtered due to their aggressive carnivorous habits towards farmers' livestock. Without a traditional predator / prey environment, the rabbits were left to an uninhibited lifestyle.

10 In the last 150 years, Australia has tried many methods to halt the spreading rabbit population. The legislature even passed laws that, among other things, required landowners to trap, poison, or kill rabbits on their property. Because of the enormous numbers of rabbits, every attempt to cut down the rabbit population was doomed to failure. After World War Ⅱ, Australia declared war on the rabbits and released a deadly virus that was 99. 8% effective at killing them. At last, Australians thought they had discovered the "magic bullet". However, the 0. 2% of the rabbit population that survived was immune to the virus. As the immune rabbits reproduced, they passed on the immunity to their offspring, and the new immune rabbit population ballooned into the millions. It seems that Australians are destined to live with millions of rabbits for quite some time.

(Australian Government Department of Sustainability, Environment, Water, Population and Communities, 2022)

New Words and Expressions

predator /ˈpredətə(r)/ *n.* 捕食性动物

prolifically /prəˈlɪfɪklɪ/ *adv.* 多产地

exponentially /ˌekspəˈnenʃəlɪ/ *adv.* 指数级地

unlikely /ʌnˈlaɪklɪ/ *adj.* 不太可能的

warren /ˈwɔrən/ *n.* 兔子窝

burrow /ˈbʌrəʊ/ *v.* 挖掘(洞或洞穴通道),挖洞

meadow /ˈmedəʊ/ *n.* 草地,牧场

ringbark /ˈrɪŋbɑːk/ *v.* 环割(树的)一圈树皮

seedling /ˈsiːdlɪŋ/ *n.* 秧苗,幼苗

bedrock /ˈbedrɔk/ *n.* 基岩

erratic /ɪˈrætɪk/ *adj.* 不规则的

larvae /ˈlɑːviː/ *n.* 幼虫,幼体

forage /ˈfɔrɪdʒ/ *vi.* (尤指动物)觅食

pollinator /ˈpɔlɪneɪtə(r)/ *n.* 传粉者

livestock /ˈlaɪvstɔk/ *n.* 牲畜,家畜

extermination /ɪkˌstɜːmɪˈneɪʃn/ *n.* 消灭,根绝

exotic /ɪgˈzɔtɪk/ *adj.* 奇异的,异国风情的

marsupial /mɑːˈsuːpɪəl/ *n.* 有袋类动物

rodent /ˈrəʊd(ə)nt/ *n.* 啮齿目动物(如老鼠等)

roam /rəʊm/ *v.* 漫游

carnivorous /kɑːˈnɪvərəs/ *adj.* 食肉的;肉食性的

uninhibited /ˌʌnɪnˈhɪbɪtɪd/ *adj.* 不受约束的

Exercises

Section I Knowledge Focus

Task 1 ***Read Paras. 1-6 the text and complete the following diagram by filling in the blanks with words from these paragraphs. (ONE WORD ONLY)***

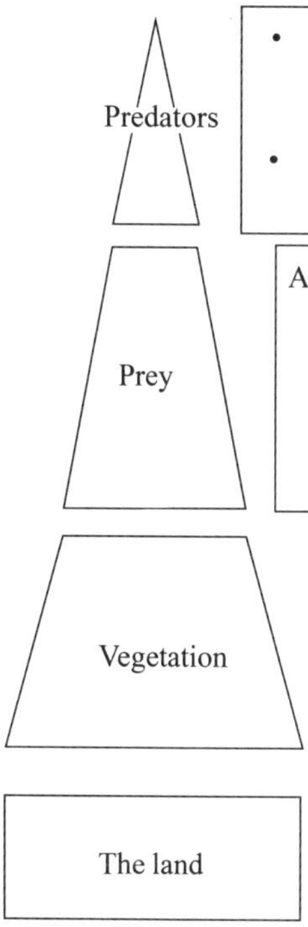 Predators	• The 1)__________ seemed to take advantage of their relatively 2)______ surroundings and turned their attention to catching easier prey including rare ground birds, indigenous marsupials, and rodents. • The 3) __________ were frequently slaughtered due to their 4)__________ carnivorous habits towards farmers' livestock
Prey	Animals going after the 5) __________ food sources as the rabbits: • the 6) __________ of several small ground-dwelling mammals of Australia's arid lands. • the 7) __________ in numbers of many native animals, such as the pig-footed bandicoot and the greater bilby. • declining populations of farm 8)__________
Vegetation	Damaging vegetation: • ringbarking trees and shrubs. • preventing 9) __________ by eating seeds and seedlings. • removing above-ground and below-ground vegetation through 10)__________ and warren construction
The land	Degrading the land: • causing erosion—the loss of 11)__________ by wind and rain. • reducing Philip Island to 12) __________ in the Norfolk Island group

Task 2 ***Read Paras. 3-4 and complete the following diagram by filling in the blanks with words from these paragraphs. (ONE WORD ONLY)***

General features	Further explanations	Facts in Australia
These rabbits are extremely 1) ______.	All the rabbits need is soil that is fit to 2) __________ and short grasses to graze on.	Since these conditions are fairly easy to come by, they can adapt to new 3) __________ such as the deserts and plains of Australia as easily as the meadows of Europe.
They are also known for rapidly producing large quantities of 4) __________.	Feral rabbits can 5) ______ from the age of four months, and can do so at any time of the year…	In favorable conditions, they can 6) __________ five or more litters in a year, with four or five young in each litter. Even in 7) __________ conditions, they can produce one or two litters a year.

Task 3 ***Given below are five statements. Each statement contains information given in one of the paragraphs of the text. Identify the paragraph from which the information is derived. Answer the questions by writing down the paragraph number (1-11) for each statement.***

__________ 1) To solve the rabbit problem, the authorities resorted to predators to prey on rabbits; however, they failed and even made things worse by introducing foxes, which turned out to be the new threat to native animals.

__________ 2) Indigenous animals were driven to starvation and even to extinction because rabbits are consuming the same food source as they do.

__________ 3) Feral rabbits can produce litters when they are very young and at any time of the year, even in the unfavorable conditions.

__________ 4) At the end of the 19th century, the number of rabbits soared from the original 24 to millions and could be found in almost all corners of Australia.

__________ 5) Wherever witchetty bushes co-exist with rabbits, they couldn't grow tall or old enough before being destroyed by rabbits, and their recovery takes several decades.

Section II Language Focus

Task 4 ***Match the term in the left column with an explanation given in the right column and write the corresponding letter in the space provided below.***

1) predator	A. *n.* a young plant that has grown from a seed
2) seedling	B. *n.* solid unweathered rock lying beneath surface deposits of soil
3) meadow	C. *n.* an animal that kills and eats other animals
4) bedrock	D. *n.* an insect that carries pollen from one flower to another
5) pollinator	E. *n.* a field covered in grass, used especially for hay

1)__________ 2)__________ 3)__________ 4)__________ 5)__________

Task 5 ***Complete the following sentences with appropriate words or expressions given below. Change the form where necessary.***

prolifically exotic erratic arid exponentially

1) The __________ fluctuation of market prices is in consequence of unstable economy.
2) Report finds debris in Earth's orbit growing __________.
3) The introduction of __________ species to new environments has often led to unintended ecological imbalances.
4) She began writing short fiction in the mid-1970s, and has continued to publish __________ to the present day.
5) In many __________ regions, the challenge of water scarcity has driven innovative

solutions in environmental management.

prey roam accommodate forage extermination

6) Stray dogs __________ the streets among collapsed houses.
7) The school was not big enough to __________ all the children.
8) Many ants are small and __________ primarily in the layer of leaves and other debris on the ground.
9) Increasing salinity caused by the evaporation resulted in the __________ of scores of invertebrate species.
10) The spider must wait for __________ to be ensnared on its web.

Task 6 ***Match the words in the left column with those in the right column to form appropriate expressions. Then complete the following sentences with one of the expressions. Change the form or add articles where necessary.***

unlikely	creature
adaptable	condition
unfavorable	area
lay	litter
produce	war
declare	eggs

1) Transparency is appearing in a number of likely and __________ of our lives.
2) The invasive species has __________ on the native forest ecosystem, causing significant disruptions to the ecological balance and biodiversity.
3) Females turtles accurately return to the same beach where they were born to __________ during breeding season.
4) Their common point is that they are not afraid of difficulties and to fight against their fates in the most __________.
5) These mice __________ size of 12 on average by the age of 7-10 months.
6) Foxes are incredibly __________ and can survive in a variety of environments.

Task 7 ***Translate the following English paragraph into Chinese.***

In the last 150 years, Australia has tried many methods to halt the spreading rabbit population. The legislature even passed laws that, among other things, required landowners to trap, poison or kill rabbits on their property. Because of the enormous numbers of rabbits, every attempt to cut down the rabbit population was doomed to failure. After World War Ⅱ, Australia declared war on the rabbits and released a deadly virus that was 99.8% effective at killing them. At last Australians thought they had discovered the "magic bullet". However, the 0.2% of the rabbit population that survived was immune to the virus. As the immune rabbits reproduced, they passed on the immunity to their offspring, and the new immune rabbit population ballooned into the millions. It seems that Australians are destined to live with millions of rabbits for quite some time.

Section III Sharing Your Ideas

Invasive species, also called introduced, alien, or exotic species, are nonnative species that can significantly modify or disrupt the ecosystems they colonize. While some species migrate naturally, human activities, such as global commerce and the pet trade, have accelerated their spread worldwide.

Task 8 ***Can you provide more examples of invasive species and the impacts they have caused? Suppose you were a leader responsible for protecting your country's ecosystems, what policies or measures would you implement to prevent further invasions?***

Part Three China's Environmental Story

Active Reading 3

Warming Up

Task ***Soybeans have diverse culinary applications, allowing the preparation of numerous edible products. List as many soybean-based foods as possible and introduce one of your favorites to your partners.***

Reading

Soybean

Terminology

1 The terms "soy" and "soya" are said to have derived from the Japanese word *shoyu* (or *sho-yu*) that designates a sauce made from salted beans. But the Japanese word may well have been inspired by the ancient Chinese name for the bean, which was *sou*. In Chinese, the word for soy sauce is *jiangyou* (or *chiang-yiu*). C. V. Piper and W. J. Morse (1923) have recorded more than 50 names for the soybean or its sauce in East Asia. In English the bean has been called soya bean, soya, soy, Chinese pea, and Japanese pea, to provide just a few of its appellations. For the purposes of this chapter, soya is used synonymously with the soybean and its many products.

Early History

2 Present-day soybean varieties (*Glycine max*), of which there are more than 20,000, can be traced to the wild soybean plant *Glycine soja* that grew in abundance in northeastern China (Hymowitz et al., 1981). Legends abound concerning the discovery and domestication of this food plant that today is the most widely used in the world (Toussaint-Samat, 1993). Around 2700 B. C., the legendary Chinese emperor Shen Nung is said to have ordered plants to be classified in terms of both food and medicinal value, and soybeans were among the five principal and sacred crops (Shih, 1959). This dating squares nicely with the judgment of modern authorities on Asian plants that soybeans have been cultivated for at least 4,500 years (Herklots, 1972). But there are other sources that indicate that the domesticated soybean (*G. max*) was introduced to China only around 1000 B. C. perhaps from the Jung people who lived in the northeast (Trager, 1995).

3 The court poems of the *Book of Odes*, from the sixth century B. C., also indicate that the

wild soybean came from northern China and that its cultivation began around the fifteenth century B. C. Confucius, who died in 479 B. C. , left behind writings that mentioned at least 44 food plants used during Chou times; they included soybeans. But they do not seem to have been very popular in ancient times. Soybeans were said to cause flatulence and were viewed mostly as a food for the poor during years of bad harvests. Nonetheless, soybeans were recorded in the first century B. C. as one of the nine staples upon which the people of China depended, and certainly there were enough people. The first official census conducted in Han China at about that time counted 60 million people, and even if such a number seems implausibly high—especially in light of a census taken in A. D. 280 that showed only 16 million—it still suggests that Chinese agricultural policies were remarkably effective, both in feeding large numbers of people and, one suspects, in encouraging the growth of large numbers of people (Chang, 1977). The famine in China in the year A. D. 194 may have been the result of too many mouths to feed and thus responsible, at least partly, for the discrepancy in the two censuses. But in addition, famine forced the price of millet to skyrocket in relation to soybeans, resulting in an increased consumption of the latter—often in the form of bean conjee or gruel (Flannery, 1969).

Early Dissemination

4 Because the wild soybean was sensitive to the amount of daylight, and because the length of growing seasons varied from region to region, domestication involved much experimental planting and breeding to match different varieties with different areas. That this was done so successfully is a tribute to ancient Chinese farmers who, as noted, were doubtless impelled by an ever-increasing need to feed larger and larger populations of humans and animals. Soybeans ground into meal and then compressed into cakes became food for travelers and soldiers on the move who, in turn, widened knowledge of the plant.

5 Buddhist priests, however, were perhaps as instrumental as anyone in the domestication of the soybean and absolutely vital to its dissemination (Yong et al. ,1974). As vegetarians, they were always interested in new foods and drinks (such as tea, which they also nurtured to an early success in China). In their monasteries, they experimented with soybean cultivation and usage and found flour, milk, curd, and sauce made from soy all welcome additions to their regimes. As missionaries, they carried the soybean wherever they went, and in the sixth century A. D. , they introduced it to Japan from Korea, which they had reached in the first century. Buddhism merged with the native Shinto religion, and the plant quickly became a staple in the Japanese diet.

6 Not only missionaries but also soldiers, merchants, and travelers helped introduce soybeans to Asian countries. The northern half of Vietnam had soybean food products as early as 200 B. C. During the sixth through the tenth centuries A. D. , Thailand received soybeans

from southwest China, and India was exposed to them during the twelfth century by traders from Pakistan.

Recent History and Dissemination

7 The Portuguese began trading in East Asia during the sixteenth century, as did the Spanish and later the Dutch. Yet the soybean was not known in Europe until the end of the seventeenth century when Engelbert Kaempfer published his *Geschichte und Beschreibung von Japan*, an account of his visit to that country during the years 1692-1694 as a guest of the Dutch East India Company. He wrote of the bean that the Japanese prized and used in so many different ways, and in 1712, he attempted, not very successfully as it turned out, to introduce this miracle plant to Europe. Its products simply did not fit into the various cuisines of the continent, which, in any event, were only then in the process of fully utilizing the relatively new American plants, such as maize and potatoes.

8 The botanists, however, were thrilled to have a new plant to study and classify, and Carolus Linnaeus, who described the soybean, gave it the name *Glycine max*. Glycine is the Greek word for "sweet", and "max" presumably refers to the large nodules on the root system, although other sources suggest that the word "max" is actually the result of a Portuguese transcription of the Persian name for the plant (Toussaint-Samat, 1992). Because of scientific interest, the soybean was shuttled about the Continent during the eighteenth century for experimental purposes.

9 In 1765, soybean seeds reached the American colonies with a sailor named Samuel Bowen, who was serving aboard an East India Company ship that had just visited China. Bowen did not return to the sea but instead acquired land in Savannah, Georgia, where he planted soybeans and processed his first crop into Chinese vetch, soy sauce, and a starchy substance incorrectly called *sago*. In North America as in Europe, however, soybean products did not go well with the various cuisines, and the bean remained little more than a curiosity until the twentieth century, despite efforts to reintroduce it.

10 By the mid-nineteenth century, the soybean was being rapidly disseminated around the globe as trade, imperialism, clipper ships, and then steamships all joined to knit the world more closely together. The expedition of Commodore Matthew Perry that opened Japan to trade in 1853-1854 returned to the United States with the "Japan pea"—actually 2 soybean varieties that were subsequently distributed by the U. S. Commissioner of Patents to farmers for planting. But lacking knowledge of and experience with the plant, the recipients were apparently not successful in its cultivation.

11 During the American Civil War, when shipping was disrupted, soybeans were frequently substituted for coffee, especially by soldiers of the Union Army (Crane, 1933). Interest also arose in soybean cultivation as a forage plant, and the Patent Office and the new Department of Agriculture (USDA) encouraged experimental planting. The USDA's role

in promoting agricultural research, regulating the industry, and serving as an information generator for farmers proved invaluable to all farmers and certainly to those growing soybeans for the first time (Arntzen et al., 1994).

(Adapted from Kiple, 2000)

New Words and Expressions

terminology /ˌtɜːmɪˈnɒlədʒɪ/ *n.* (某学科的)术语

designate /ˈdezɪɡneɪt/ *vt.* 命名;指定

appellation /ˌæpəˈleɪʃ(ə)n/ *n.* 名称,称呼,称号

abound /əˈbaʊnd/ *vi.* 大量存在,有许多

dating /ˈdeɪtɪŋ/ *n.* 年代测定

flatulence /ˈflætjʊləns/ *n.* 肠胃胀气

implausibly /ɪmˈplɔːzəblɪ/ *adv.* 难以置信地

discrepancy /dɪˈskrepənsɪ/ *n.* 差异,不符合,不一致

conjee /ˈkɒndʒiː/ *n.* (等于 congee)粥,稀饭

gruel /ˈgruːəl/ *n.* 稀粥

dissemination /dɪˌsemɪˈneɪʃ(ə)n/ *n.* 散播

tribute /ˈtrɪbjuːt/ *n.* (良好效果或影响的)体现,显示

impel /ɪmˈpel/ *vt.* 推动,驱使

grind /graɪnd/ *v.* 磨碎,碾碎

instrumental /ˌɪnstrʊˈment(ə)l/ *adj.* 起重要作用的

monastery /ˈmɒnəst(ə)rɪ/ *n.* 修道院;寺院

regime /reɪˈʒiːm/ *n.* 养生法

botanist /ˈbɒtənɪst/ *n.* 植物学家

nodule /ˈnɒdjuːl/ *n.* (尤指植物上的)节结,小瘤

starchy /ˈstɑːtʃɪ/ *adj.* 含大量淀粉的

disseminate /dɪˈsemɪneɪt/ *vt.* 散布,传播(信息、知识等)

commodore /ˈkɒmədɔː(r)/ *n.* 海军准将

commissioner /kəˈmɪʃənə(r)/ *n.* (政府部门等的)重要官员

recipient /rɪˈsɪpɪənt/ *n.* 接受方;领受者

Exercises

Section I　Understanding the Text

Task 1 ***Discuss the following questions in small groups.***

1) What is considered to be the origin of present-day soybean varieties?

2) Why weren't the soybeans very popular in ancient times of China?

3) Why did the year A. D. 194 see an increase of consumption of the soybeans in China?

4) Why did the domestication of soybeans involve much experimental planting and breeding?

5) Why were Buddhist priests considered by scholars as instrumental in the domestication of the soybean and vital to its dissemination?

6) Who contributed to the dissemination of soybeans to Asian countries?

7) Why was the soybean spread in Europe during the 18th century?

8) Why did farmers start to grow soybeans in America?

Section II Developing Critical Thinking

Biosphere reserves are essential for promoting sustainability, preserving biodiversity, and balancing human-environment relations. The East Asian Biosphere Reserve Network (EABRN), established in 1994, includes countries like China, Japan, and Russia. These reserves contribute to the World Network of Biosphere Reserves under UNESCO's Man and the Biosphere (MAB) Program. China, with 34 biosphere reserves, plays a key role in this global initiative.

Task 2 ***As one of China's biosphere reserves, Mount Huangshan, located in Anhui Province, requires a periodic review to evaluate its conservation, research, and sustainable development efforts. This review will assess the reserve's ecological health, local community involvement, and alignment with EABRN and UNESCO goals. To write it, you can follow the instructions below.***

Steps for writing a periodic review

Step	Description
1) Define the purpose and scope	Clearly outline the objectives of the review and define the time period and focus areas to be assessed
2) Gather relevant data and information	Collect necessary data, reports, research findings, and feedback to assess the subject comprehensively
3) Analyze strengths and weaknesses	Identify what is working well and areas that need improvement based on the evaluation
4) Provide insights and analysis	Present an analysis explaining the reasons behind successes or shortcomings and link findings to broader goals
5) Make recommendations	Provide actionable, practical recommendations for improvement or future actions
6) Conclusion	Summarize the main findings and reiterate key recommendations and next steps

Typical language elements of a periodic review

1) Objective and analytical tone: The language is neutral, factual, and focused on analysis rather than opinion. It presents information based on evidence and data, avoiding emotional or biased language.
2) Use of data and evidence: Statistical data, research findings, and concrete examples are frequently incorporated to support assessments and conclusions. Phrases like "according to the data", "research indicates", or "evidence suggests" are common.
3) Formal and professional language: The review avoids colloquialisms and casual expressions. Instead, it uses formal vocabulary to convey authority and professionalism, such as "assess", "evaluate", "examine", and "recommend".

4) Use of passive voice: Passive voice are often used to focus on actions rather than the actors themselves, such as "data was collected" or "the reserve was established".
5) Comparative language: When evaluating performance or outcomes, comparative phrases may be used, such as "compared to previous years", "in contrast with", or "relative to".
6) Recommendations and actionable language: The review ends with specific, actionable recommendations using language such as "It is recommended that...", "Future efforts should focus on...", or "Improvement is needed in...".

UNIT 5

Industrialization and Environment

Part One Preparation

Unit Preview

Since humans entered the industrial age, air pollution has accompanied their lives and production. Before the discovery of coal, the main energy sources were wood, manpower, and wind. The emergence of coal replaced wood, breaking through the energy bottleneck for industrialization. Being the pioneer of the Industrial Revolution, Britain utilized the energy provided by its coal reserves to establish the world's most technologically advanced, dynamic, and prosperous economy from 1780 to 1880.

The flames of burning coal and towering chimneys have become symbols of Britain's industrialization. However, it was the widespread use of coal that ultimately led to the most severe period of air pollution in British history. The rolling smoke of harmful substances released during coal combustion, such as sulfur dioxide, had a direct impact on the environment. London was the worst city for air pollution. The mixture of smoke and fog which turned yellow and black, often shrouded London for many days without dispersing. In 1905, the term "smog" was coined in English, specifically referring to the mixture of fog and coal smoke or coal dust.

Since humans entered the industrial age, air pollution has accompanied their lives and production. Before the discovery of coal, the main energy sources were wood, manpower, and wind. The emergence of coal replaced wood, breaking through the energy bottleneck for industrialization.

Being the pioneer of the Industrial Revolution, Britain utilized the energy provided by its coal reserves to establish the world's most technologically advanced, dynamic, and prosperous economy from 1780 to 1880.

The flames of burning coal and towering chimneys have become symbols of Britain's industrialization. However, it was the widespread use of coal that ultimately led to the most severe period of air pollution in British history. The rolling smoke of harmful substances released during coal combustion, such as sulfur dioxide, had a direct impact on the environment. London was the worst city for air pollution. The mixture of smoke and fog which turned yellow and black, often shrouded London for many days without dispersing. In 1905, the term "smog" was coined in English, specifically referring to the mixture of fog and coal smoke or coal dust.

As a latecomer to industrialization, China prioritizes sustainable development, learning from past mistakes. Efforts to recycle industrial waste, such as ceramics, into eco-friendly materials reflect its commitment to balancing economic growth with environmental protection.

Learning Objectives

Upon completion of this unit, you will be able to:

- gain a better understanding of various environmental pollutants.
- analyze the impact of the industrialization on the environment.
- write about the relationship between industrialization and environmental deterioration.

Part Two Global Perspectives

Active Reading 1

Warming Up

Task 1 ***Try to answer the following questions.***

1) Have you heard of the "London Smog Incident" that occurred in London from December 5 to 9, 1952?

2) What do you know about this incident?

Task 2 ***Watch a short video clip. Then discuss the following questions with your partners.***

视听资源

1) What happened in London in December of 1952?

2) What was the cause of the tragedy, according to scientists?

Reading

Atmospheric Pollution in London

1 Smoke and winter fogs were the most obvious signs of atmospheric pollution. In some towns, and parts of towns, there were other pollutants as unpleasant and sometimes more dangerous to health. They can be summed up as being in one of three classes: acids, dust, and a miscellaneous collection of bad smells—effluvia, a ponderous and euphemistic word preferred by the Victorians. The acid emissions that arose from the burning of coal were much the largest. Carbon, thoroughly burnt, turns to carbon dioxide, and large and growing quantities of this gas were emitted in the nineteenth century. It aroused little concern at the time since carbon dioxide does not dissolve very readily in water to form carbonic acid, and the acid is in any case weak. A handful of scientists—Fourier in the 1820s and Arrhenius in 1896—pointed to the dangers of the greenhouse effect, but this problem aroused little interest before the late 1980s. Sulphur is always present in coal, and usually accounts for between one and two per cent of total weight. Like carbon dioxide, sulphur dioxide, and sulphur trioxide do not combine so freely with water as to deposit sulphurous and sulphuric acid in large doses close to where coal has been burnt. As recent experience of acid rain has shown, high chimneys successfully distribute sulphur compounds over a very wide area. Nevertheless the damage to masonry and ironwork from corrosive acids in the air was obvious in nineteenth-century towns and the acrid taste of winter fogs was an unpleasant experience. Nobody in authority saw fit to protect the public from the acids arising from the combustion of coal. Acid emitted in particular industrial

processes was another matter.

2 The most famous example of industrial pollution by the emission of acid occurred in the alkali manufacture, the largest branch of the heavy chemical industry in nineteenth-century Britain. The principal product of the alkali manufacture was soda, or sodium carbonate. Soda was valuable in glass-making, and as a cleaning agent in its own right; and caustic soda, a derivative, was an ingredient of soap. Put simply, the manufacture of alkali (soda) by the Leblanc Process took place in two stages: first, salt combined with sulphuric acid to produce sodium sulphate and an unwanted byproduct—hydrochloric acid. Sodium sulphate was later burnt with chalk and coal to produce soda and another troublesome byproduct—calcium sulphide. Of the two byproducts, hydrochloric acid was by far the more pernicious. It was given off from open furnaces as a gas and since it readily dissolved in water, it was little diluted before it came to earth. The atmosphere in towns like Runcorn was distinctly sharp as recently as the early 1950s when alkali manufacture was much reduced. In the nineteenth century Runcorn, Widnes, and St. Helens, where the alkali manufacture had settled, were notorious for the impurity of their air. For several miles on the leeward side of the alkali works trees died and crops and grass grew poorly. There were other important centres of the industry in Glasgow and on Tyneside, and some smaller works in London, Bristol and elsewhere.

3 In 1836, William Gossage, whose works then lay at Stoke Prior in Worcestershire, patented a device for condensing the hydrochloric acid in towers packed with brushwood (later coke was used) down which water trickled, absorbing the gas and producing a weakish solution of acid. If most manufacturers had adopted this process, and if they had then found a use for the hydrochloric acid, the atmosphere on Merseyside and elsewhere would have been sweeter. Unhappily Gossage's towers were not much seen in the industrial landscape and when they were installed the acid was often discharged into the nearest river. Not only did the acid further pollute some already unwholesome waters, it also contributed to another form of atmospheric pollution. This arose from the combination of hydrochloric acid (either raining down from the skies or discharged into rivers) with the calcium sulphide that also resulted from alkali manufacture. The combination of these two chemicals released sulphuretted hydrogen. The smell (like rotten eggs) sometimes spread far and wide over south-west Lancashire. The nuisance became so bad that in 1876 a royal commission was appointed to report on this and other noxious vapours. A Liverpool solicitor P. F. Garnett living at Aigburth, seven miles from Widnes, kept a record of stenches arising from alkali waste. He noticed the stench 24 times in 1875 and 29 times in 1876. On one day in April of that year it was perceptible in Castle Street, Liverpool, some twelve miles from its point of origin. Until a use could be found for the hydrochloric acid, the solid alkali waste, or both, there were only poor prospects of ending this particularly

unpleasant kind of atmospheric pollution.

4 Though the most extensive and the best-known source of pollution by hydrochloric acid, alkali manufacture did not stand alone. Wherever salt or hydrochloric acid was used in industry, the danger of unpleasant emissions arose. The rich salt deposits of Cheshire have long been a mainstay of the British chemical industry, and many salt-pans were built to evaporate the brine pumped up from beneath the Cheshire plain. Not all the salt had to be pumped, for there were also enormous deposits of rock salt to be mined in the usual way. But brine was important and in the heyday of the salt pans more than 1,000 were at work. Unless carefully controlled, the process gave off substantial quantities of hydrochloric acid gas, admittedly in the neighbourhood of relatively small Cheshire towns like Northwich and Middlewich, where fewer people suffered annoyance than from the alkali manufacture on Merseyside.

5 Another source of emissions was the Yorkshire shoddy trade. The shoddy trade reprocessed woollen rags by tearing them to shreds and spinning the product into an inferior yarn. This yarn, either alone or combined with yarn made from new wool, was then woven into cloth, not all of it deserving the name of shoddy. Problems arose when the rags were mixture fabrics containing cotton or linen as well as wool. The vegetable fibres stubbornly refused to submit to reworking and had to be eliminated. Luckily they would dissolve in a weak solution of sulphuric acid or in hydrochloric acid gas, leaving the wool unscathed. This process—known as carbonisation because it charred the unwanted fibres—was introduced in the early 1850s, not only in the shoddy districts of the West Riding, but also in London, the centre of the trade in rags. The product, known as extract, was disliked in the shoddy trade because its felting qualities were inferior to those of yarn made from pure woollen rags, and a large part of the output found its way to foreign mills. The manufacturers, until advised by government inspectors, made little effort to condense the hydrochloric acid gas and in 1884, shortly after inspection had begun, only one works had a condensing tower.

(Adapted from Clapp, 1994)

New Words and Expressions

miscellaneous /ˌmɪsəˈleɪnɪəs/ *adj.* 混杂的，各种各样的

effluvium /eˈfluːvɪəm/ *n.* 恶臭，臭气

ponderous /ˈpɒndərəs/ *adj.* （言语或文字）慢条斯理的；沉闷乏味的

euphemistic /juːfɪˈmɪstɪk/ *adj.* 委婉的

dissolve /dɪˈzɒlv/ *v.* 溶解；使溶解

carbonic /kɑːˈbɒnɪk/ *adj.* 含碳的

sulphur /ˈsʌlfə/ *n.* 硫；硫黄

deposit /dɪˈpɒzɪt/ *vt.* （尤指河流或液体）使沉积，使沉淀，使淤积

masonry / ˈmeɪsənrɪ/ *n.* 砖石结构，砖石建筑

corrosive /kəˈrəʊsɪv/ *adj.* 腐蚀性的，侵蚀

性的

acrid /ˈækrɪd/ *adj.* （气、味）辛辣的，难闻的，刺激的

combustion /kəmˈbʌstʃən/ *n.* 燃烧

alkali /ˈælkəˌlaɪ/ *n.* 碱

sodium /ˈsəʊdɪəm/ *n.* 钠

carbonate /ˈkɑːbəˌneɪt/ *n.* 碳酸盐

agent /ˈeɪdʒənt/ *n.* （化学）剂

caustic /ˈkɔːstɪk/ *adj.* 腐蚀性的

sulphate /ˈsʌlfeɪt/ *n.* 硫酸盐

hydrochloric /ˈhaɪdrəˈklɒrɪk/ *adj.* 氯化氢的；盐酸的

calcium sulphide /ˈkælsɪəm ˈsʌlfaɪd/ *n.* 硫化钙

pernicious /pəˈnɪʃəs/ *adj.* 有害的，恶性的

dilute /daɪˈluːt/ *vt.* 稀释，冲淡

leeward /ˈliːwəd/ *adj.* 在背风面的；背风的；下风的

trickle /ˈtrɪkəl/ *v.* （使）滴，淌，小股流淌

solution /səˈluːʃən/ *n.* 溶液

unwholesome /ʌnˈhəʊlsəm/ *adj.* 有损健康的，不健康的，不卫生的

sulphuretted hydrogen /ˈsʌlfjʊˌretɪd haɪdrədʒən/ *n.* 硫化氢

solicitor /səˈlɪsɪtə/ *n.* 事务律师，诉状律师（代拟法律文书、提供法律咨询等的一般辩护律师）

stench /stentʃ/ *n.* 臭气，恶臭

mainstay /ˈmeɪnˌsteɪ/ *n.* 支柱，中流砥柱

evaporate /ɪˈvæpəˌreɪt/ *v.* （使）蒸发，挥发

brine /braɪn/ *n.* 盐水

shoddy /ˈʃɒdɪ/ *adj.* 做工粗糙的，粗制滥造的，劣质的

unscathed /ʌnˈskeɪðd/ *adj.* 未受伤的，未受伤害的

char /tʃɑː/ *v.* （使）烧黑，烧焦

condense /kənˈdens/ *v.* （由气体）冷凝，（使气体）凝结

condensing tower /kənˈdensɪŋ ˈtaʊə/ *n.* 冷凝塔

Exercises

Section I　Knowledge Focus

Task 1 ***Read the whole passage and complete the following diagram by filling in the blanks with words from the passage.*** （*NO MORE THAN TWO WORDS*）

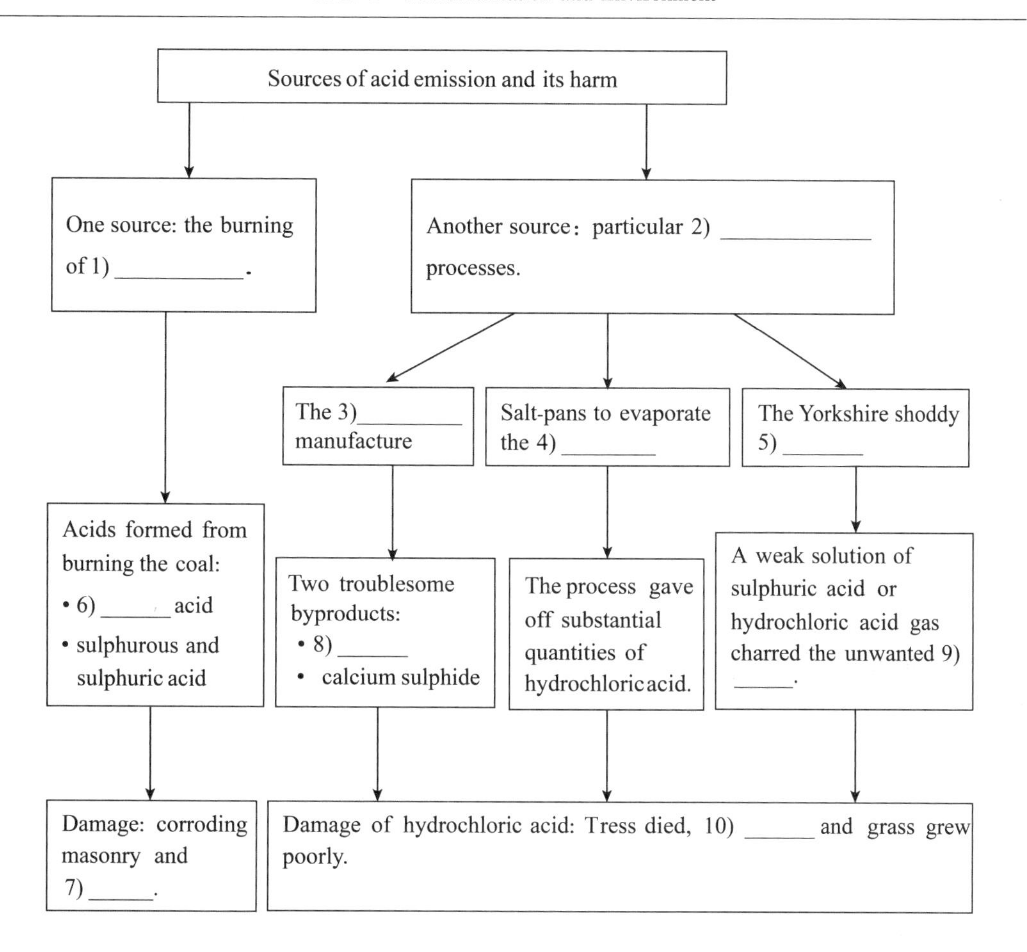

Task 2 ***Read the passage and decide whether the following statements are true or false. Write T for true or F for false in brackets in the blank provided before each statement.***

__________ 1) In the 19th century, researchers discovered that large amounts of acid emission from coal combustion became the main air pollutants in Britain at that time.

__________ 2) The manufacture of alkali by the Leblanc Process consisted of two stages. First, salt was combined with sulphuric acid to produce sodium sulphate, and then sodium sulphate was burnt with chalk and coal to produce soda.

__________ 3) Gossage's towers for condensing the hydrochloric acid were not commonly accepted in industrial areas of Britain. When they were installed, atmosphere improved, but at the same time, another pollutant, sulphuretted hydrogen, might also be released into the air.

__________ 4) Ending the unpleasant kind of atmospheric pollution caused by hydrochloric acid depended partly on finding a way to make use of it.

__________ 5) By 1884, almost no manufacturers in Britain used condensing towers to treat hydrochloric acid gas.

Section II Language Focus

Task 3 ***Identify what is being described in the following phrases. Choose your answer from the box below and write the corresponding word in the space provided.***

miscellaneous emission condense fabric fibre

__________ 1) to change from a gas into a liquid; to make a gas change into a liquid

__________ 2) consisting of many different kinds of things that are not connected and do not easily form a group

__________ 3) one of the many thin threads that form body tissue, such as muscle, and natural materials, such as wood and cotton

__________ 4) production or sending out of light, heat, gas, etc.

__________ 5) material made by weaving wool, cotton, silk, etc., used for making clothes, curtains, etc. and for covering furniture

Task 4 ***Complete the following sentences with appropriate words or expressions given below. Change the form where necessary.***

notorious pollutant dissolve trickle agent

1) Industrial __________ are responsible for a sizable proportion of many cancers.
2) The bar has become __________ as a meeting place for drug dealers.
3) Common oxidizing __________ are oxygen, hydrogen peroxide and ferric salts.
4) A tear __________ down the old man's cheek.
5) We have to find a way to __________ the paint on the furniture and uncover its original.

solution evaporate discharge mainstay noxious

6) The factory was fined for __________ chemicals into the river.
7) Many household products give off __________ fumes.
8) Farming is still the __________ of economy in this country.
9) A __________ of salt water for example, will have a different freezing point depending on how much salt is dissolved in the water.
10) Since alcohol makes up nearly half of the vodka, it will __________ as it cooks.

Task 5 ***Match the words in the left column with those in the right column to form appropriate expressions. Then complete the following sentences with one of the expressions. Change the form or add articles where necessary.***

corrosive	interest
distinctly	national identity
stand	impact
arouse	alone
dilute	different

1) Tracing those lines to genes underlying disease is hard work, but the researchers believe that this may be the only way to __________ in their project.
2) Many citizens in Britain complained that this policy would __________ and hurt standards of living.
3) This has had a "__________ on the UK economy", the report concludes, as household spending power has shrunk.
4) The two nations have __________ stances on carbon dioxide emissions and environmental quality.
5) The case of Britain having accelerated industrialization at the expense of environment does not __________.

Task 6 ***Complete the following sentences with the correct form of the words in brackets.***

1) Large and growing quantities of carbon dioxide were __________(emission) in London in the 19th century.
2) In the 19th century Runcorn, Widnes and St. Helens, where the alkali manufacture had settled, were notorious for the __________ (pure) of their air.
3) The __________ (combine) of these two chemicals released sulphretted hydrogen.
4) Carbon dioxide does not dissolve very __________ (ready) in water to form carbonic acid.
5) The process, known as carbonization because it charred the __________ (want) fibres, was introduced in the early 1850s.

Task 7 ***Translate the following Chinese expressions into English with what you have learned from the text.***

1) Apart from smoke and winter fogs, there were other pollutants, which can be __________(概括，总结) as being in one of three classes: acids, dust, and a collection of bad smells.
2) It __________(几乎未引起关注) at the time since carbon dioxide does not dissolve very readily in water to form carbonic acid, and the acid is in any case weak.
3) Nobody in authority __________ (认为恰当;愿意做) protect the public from the acids arising from the combustion of coal.
4) Not only did the acid further pollute some already unwholesome waters, it also __________(成为原因) another form of atmospheric pollution.

Section III Sharing Your Ideas

Britain was both a pioneer of the Industrial Revolution and the first victim of the pollution catastrophe caused by industrial development. Do you know why London used to be called the "City of Fog"? The London Smog Incident of 1952 forced the British government to re-examine the harm caused by air pollution.

Task 8 ***Discuss with your partners what measures Britain took to address this issue.***

Active Reading 2

Warming Up

Task 1 ***George Perkins Marsh was a renowned American scholar and naturalist. Try to say something about him and his book.***

视听资源

Task 2 ***What was the origin of the Industrial Revolution? Watch a video clip and discuss the following questions with your partners.***

1) What contributed to Britain's industrialization according to some historians?

2) What factors may have spurred industrialization in Britain?

3) What kind of power did the Industrial Revolution help Britain to become?

Reading

George Perkins Marsh Spurs Consideration of Industrialization

1 In each romanticized approach to nature, little criticism was leveled at nineteenth-century life. Primarily, romantics celebrated nature while not detracting from wealthy consumers. However, as nature's beauty was seen worthy of celebration and reverence, some observers were growing increasingly unwilling to overlook the abuses wrought on it by the insensitive. Leading this vocal criticism, George Perkins Marsh emerged in the 1860s as one of the only spokesmen for a scientific orientation who was willing to take on the culture of industrialization. With the growth of cities in the United States during the nineteenth century, there was a dramatic increase in industry, and, as industry grew, the natural environment was adversely impacted in immediately visible ways. For example, the machinery of many factories was fueled by coal that caused smokestacks to belch black smoke into the air, and industrial byproducts flowed into the waterways, leaving them polluted. The impact of these industries did not go unnoticed to young Marsh, who wrote the following in 1849:

I spent my early life almost literally in the woods; a large portion of the territory of Vermont was, within my recollection, covered with natural forests; and having been personally engaged to a considerable extent in clearing lands, and manufacturing, and dealing in lumber, I have had occasion both to observe and to feel the effects resulting from an injudicious system of managing woodlands and the products of the forest.

2 A trained geographer, George Marsh still had influences that could instill in him a romantic view of the natural world. For instance, his cousin, the philosopher and University of Vermont president, James Marsh, helped to redefine transcendentalism. Instead of the

idealism of much of New England transcendentalism, with its interest in conservation or primitivism, Marsh's view of transcendentalism advocated taming wilderness. He advocated for practical informed decisions and increased command over nature. Concerning human use of natural resources, he felt that it was important to weigh the results and act accordingly.

3 Seeing the damage to the natural environment occur right before their eyes, some people became alarmed and began to search for ways to create a balance between industrial progress and the preservation of natural resources. These very early "conservationists" included George Perkins Marsh, who wrote *Man and Nature*. Marsh argued that the growth of industry was upsetting the natural balance of nature. The scale and scope of this action overwhelmed knowledgeable observers such as Vermont statesman Marsh. While acknowledging the need for human use of the natural environment, Marsh used his book *Man and Nature* to take Americans to task for their misuse and mismanagement of their national bounty. Marsh wrote the following:

Nature, left undisturbed, so fashions her territory as to give it almost unchanging permanence of form, outline, and proportion, except when shattered by geologic convulsions. ... In countries untrodden by man, the proportions and relative positions of land and water ... are subject to change only from geological influences so slow in their operation that the geographical conditions may be regarded as constant and immutable. Man has too long forgotten that the earth was given to him for usufruct alone, not for consumption, still less for profligate waste. ... But she has left it within the power of man irreparably to derange the combinations of inorganic matter and of organic life ... man is everywhere a disturbing agent. Wherever he plants his foot, the harmonies of nature are turned to discords ... of all organic beings, man alone is to be regarded as essentially a destructive power.

4 To reach his conclusions in *Man and Nature*, Marsh drew from his observations as a youth in Vermont, as well as those from travels in the Middle East. The philosophies expressed above, such as referring to humans as "disturbing agents", contradicted the conventional ideas of the time. In geography, for instance, the work of scholars including Arnold Guyot and Carl Ritter argued that the physical aspects of the earth were entirely the result of natural phenomena, such as mountains, rivers, and oceans. To suggest that humans could disrupt and, ultimately, manipulate these forms and patterns was profound. Marsh was the first to describe the interdependence of environmental and social relationships. Lowenthal wrote, "Like Darwin's *Origin of Species*, Marsh's *Man and Nature* marked the inception of a truly modern way of looking at the world. Marsh's ominous warnings inspired reforestation, watershed management, soil conservation, and nature protection in his day and ours".

5 In addition to constructing this intellectual framework for future generations, Marsh used various occupations to influence approaches to land use, including lawyer, newspaper editor, sheep farmer, mill owner, lecturer, politician and diplomat. As a congressman in Washington, Marsh helped to found and guide the Smithsonian Institution. He served as U. S. Minister to Turkey for five years, where he aided revolutionary refugees, and spent the last twenty-one years of his life (1861-1882) as U. S. Minister to the newly formed United Kingdom of Italy.

6 One of the lasting influences of his thought was to celebrate and, eventually, to preserve the remaining unspoiled places. Years of living in the Middle East afforded him time to travel throughout Egypt and part of Arabia. On one of these journeys, he developed an obsession with the camel and was convinced that the animal might thrive in the American deserts. In addition to transportation, Marsh thought that the camel could prove useful in wars in the Southwest. Inspired by a lecture Marsh delivered at the Smithsonian on his return to the States, Congress ordered seventy-four camels from the Middle East to be shipped to Texas in 1856. The experiment failed, mostly because of the onset of the Civil War and the unfamiliarity with the ways of the camel on the part of the army's equestrian division.

7 Regardless, however, Marsh brought an alternative paradigm to ideas of development and land use.

(Adapted from Black et al., 2008)

New Words and Expressions

level **/ˈlevəl/** *vt.* 使(批评,指控,控告)针对

insensitive **/ɪnˈsensɪtɪv/** *adj.* (对变化)懵然不知的,麻木不仁的

orientation **/ˌɔːrɪenˈteɪʃən/** *n.* (个人的)基本信仰,态度,观点

fuel **/fjʊəl/** *vt.* 给……提供燃料

smokestack **/ˈsməʊkˌstæk/** *n.* (工厂的)大烟囱

belch **/beltʃ/** *v.* (大量)喷出,吐出

injudicious **/ˌɪndʒʊˈdɪʃəs/** *adj.* 不明智的;不当的

instill **/ɪnˈstɪl/** *vt.* 徐徐滴入,逐渐灌输

transcendentalism **/ˌtrænsenˈdentəˌlɪzəm/** *n.* (爱默生)超验主义

primitivism **/ˈprɪmɪtɪˌvɪzəm/** *n.* (哲学、艺术或文学的)原始主义,尚古主义;原始风格

bounty **/ˈbaʊntɪ/** *n.* 慷慨之举;大量给予之物

fashion **/ˈfæʃən/** *vt.* (尤指用手工)制作,使成型,塑造

convulsion **/kənˈvʌlʃən/** *n.* 震动,震撼

immutable **/ɪˈmjuːtəbəl/** *adj.* 不可改变的,永恒不变的

usufruct **/ˈjuːsjʊˌfrʌkt/** *n.* 使用收益权

profligate **/ˈprɒflɪgɪt/** *adj.* 挥霍的,浪费的

irreparably **/ɪˈrepərəbəlɪ/** *adv.* 无法弥补

地,不能修复地,不可恢复地

derange /dɪˈreɪndʒ/ *vt.* 打乱,搅乱

discord /ˈdɪskɔːd/ *n.* 不一致;不和,纷争

inception /ɪnˈsepʃən/ *n.* (机构、组织等的)开端,创始

ominous /ˈɒmɪnəs/ *adj.* 预兆的;恶兆的,不吉利的

equestrian /ɪˈkwestrɪən/ *adj.* 马术的

paradigm /ˈpærəˌdaɪm/ *n.* 典范,范例,样式

Exercises

Section I Knowledge Focus

Task 1 ***George Perkins Marsh is a conservationist of natural resources. His views on resource conservation, species value, and the relationship between humans and nature have had a significant influence in the United States. Complete the following diagram about Marsh's ideas and his book by filling in the blanks with words from the passage.*** (***NO MORE THAN TWO WORDS***)

Problem: Industrialization and environmental damage

- 1) _____ growth of industry in the 19th century.
- 2) _____ impact on the environment: black smoke, polluted waterways, and deforestation.
- 3) Ignorance of the long-term influence on nature.

↓

Marsh's critique and advocacy

- **Critique**
  - He argued that the industrial growth was 3) _____ nature's balance.
  - He criticized Americans for misusing and 4)_____ their national resources.
  - He considered humans as "5) ____ agents". "Wherever he plants his foot, the harmonies of nature are turned to 6) _____."
- **Advocacy**
  - He championed 7) _____ wilderness, while 8) _____ the results and acting 9) _____.
  - He urged sustainable resource use: Earth was "for usufruct alone, not for 10) _____, " nor for waste.

↓

Marsh's influence and legacy

- Like Darwin's *Origin of Species*, *Man and Nature* marked the 11) _____ of a truly modern way of looking at the world.
- He inspired conservation movements (e.g. reforestation, watershed management, 12) ______, and nature protection) in his day and ours.
- People have learned to preserve the remaining 13) _____ places.

Task 2 ***Answer the following questions according to the passage.***

1) How was the natural environment in the United States affected by the dramatic increase in industry during the 19th century? Can you list some examples?

2) How did Marsh's view of transcendentalism differ from that of New England?

3) What did the very early conservationists in the United States do when faced with the damage to the natural environment?

4) Why did Marsh argue that "man alone is to be regarded as essentially a destructive power"?

5) What were Marsh's ideas about environmental protection according to the text?

Section II Language Focus

Task 3 ***Identify what is being described in the following phrases. Choose your answer from the box below and write the corresponding word in the space provided.***

accordingly	contradict	reverence	orientation	onset

__________ 1) a person's basic beliefs or feelings about a particular subject

__________ 2) the beginning of sth, especially sth unpleasant

__________ 3) in a way that is appropriate to what has been done or said in a particular situation

__________ 4) a feeling of great respect or admiration

__________ 5) to be so different from each other that one of them must be wrong

Task 4 ***Complete the following sentences with appropriate words or expressions given below. Change the form where necessary.***

advocate	upset	dramatic	ultimately	fuel

1) The drought is expected to have __________ effects on the California economy.

2) Uranium is used to __________ nuclear plants.

3) The report __________ that all buildings be fitted with smoke detectors.

4) These latest rumors could __________ the peace negotiation which are due to begin next week.

5) Whatever the scientists __________ conclude, all of their data will immediately be disputed.

obsession	shatter	ominous	thrive	proportion

6) Loam is a soil with roughly equal __________ of clay, sand and silt.

7) He dropped the vase and it __________ into pieces on the floor.

8) She picked up the phone but there was an __________ silence at the other end.

9) This south-facing slope gets day-long sunshine, and even at 1,700 metres up, grapes can still __________.

10) She turned a full 180 degrees and developed an __________ with those soap operas.

Task 5 ***Find the words /phrases in the box that have the same meaning as the underlined words or phrases in each of the following sentences and write the corresponding word or phrase in the space provided. Change the form where necessary.***

unchangeable	take away from	disturb	aim	control

__________ 1) In each romanticized approach to nature, little criticism was leveled at nineteenth-century life.

__________ 2) Primarily, romantics celebrated nature while not detracting from wealthy consumers.

__________ 3) He advocated for practical informed decisions and increased command over nature.

__________ 4) In countries untrodden by man, the proportions and relative positions of land and water.... are subject to change only from geological influences so slow in their operation that the geographical conditions may be regarded as constant and immutable.

__________ 5) But she has left it within the power of man irreparable to derange the combinations of inorganic matter and of organic life.

Task 6 ***Translate the following Chinese expressions into English with what you have learned from the text.***

1) 直言不讳的批评
2) 工业副产品
3) 自然资源
4) 得出结论
5) 自然现象
6) 环境与社会关系的相互依存

Task 7 ***Translate the following sentences into Chinese.***

1) However, as nature's beauty was seen worthy of celebration and reverence, some observers were growing increasingly unwilling to overlook the abuses wrought on it by the insensitive.
2) With the growth of cities in the United States during the nineteenth century, there was a dramatic increase in industry, and, as industry grew, the natural environment was adversely impacted in immediately visible ways.
3) Instead of the idealism of much of New England transcendentalism, with its interest in conservation or primitivism, Marsh's view of transcendentalism advocated taming wilderness.
4) Marsh argued that the growth of industry was upsetting the natural balance of nature. The scale and scope of this action overwhelmed knowledgeable observers such as Vermont statesman Marsh.
5) While acknowledging the need for human use of the natural environment, Marsh used his book *Man and Nature* to take Americans to task for their misuse and mismanagement of their national bounty.

Section III Sharing Your Ideas

Man and Nature, George Marsh's masterpiece, contains rich ideas on forest conservation. Marsh pointed out that humans and forests interact with each other.

Task 8 ***Discuss with your partners the interaction between humans and forests, provide examples to show how forest destruction impacts humans, and discuss China's achievements in forest conservation.***

Part Three China's Environmental Story

Active Reading 3

Warming Up

Task ***We use ceramics every day—plates, tiles, sinks. But what happens when they break? Have you ever thought about where all that waste goes? What do you think should be done with it?***

Reading

Recycled Products Shatter Old Ideas about Ceramic Waste

1 China has been the world's main producer of traditional ceramics for many years. In 2017, China produced approximately 10 billion square meters of architectural ceramics, 220 million pieces of sanitary ceramics, and 50 billion pieces of daily-use ceramics. The Central China region accounted for 25 percent, 55 percent, and 45 percent of these outputs respectively, while Jiangxi was the third-ranked province in traditional ceramic production. However, the thriving ceramic industry has generated a massive amount of non-degradable ceramic waste. For instance, Jingdezhen, a historic ceramic hub, produces over 60,000 metric tons of waste annually, while the nationwide total is estimated at 18 million tons.

2 Ceramic waste comes in various forms, including traditional daily-use, architectural, and sanitary ceramics, among others. Currently, some ceramic waste is recycled, but many companies still resort to traditional landfilling.

3 In the past, wood-fired kilns had a success rate of only 30 percent in ceramic production, and failures were common. As a result, a significant amount of ceramic waste was dumped as landfill into low-lying areas of Jingdezhen to prevent flooding and waterlogging. This led to the development of densely populated areas with extensive construction after hills and streams were leveled and covered. Ceramic waste and kiln bricks were also used to construct Jingdezhen's sewers. Even today, walls from the Qing Dynasty (1644–1911) and pools from the Ming Dynasty (1368–1644), built with discarded ceramics, kiln bricks, and other fragments, can be seen among the ruins of the Imperial Porcelain Factory in Jingdezhen.

4 However, as industrial production has grown, this traditional approach has become unsustainable. Ceramics do not degrade, occupying vast land areas when buried, and some glazed waste releases harmful chemicals over time, threatening air quality and groundwater.

Waste into Wealth

5 In recent years, China has placed greater emphasis on environmental issues, prompting a shift in waste management practices.

6 A notable example is a 2021 recycling initiative in Jingdezhen, launched by Yi Design, a materials company, and the Letian Pottery Workshop. Local ceramic artists and vendors now collect waste after production and deliver it to designated recycling points, signaling a growing environmental consciousness.

Ceramic "mountains" Cleared

7 In Chaozhou, Guangdong province, another major ceramic production city in China, steps have been taken in recent years to clean up waste problems. Due to a wastage rate of 5 to 10 percent during the production process, the accumulation of "ceramic mountains" posed a significant problem for the city's ecology. In response, local authorities established a waste disposal site in 2017 and introduced special operations to address the problem. They also encouraged larger firms to upgrade their equipment and technology, and use market mechanisms to achieve breakthroughs in comprehensive recycling. Within a year, Chaozhou had processed 67,000 tons of ceramic waste.

8 Nationwide policies have further supported these efforts. In 2018, the Ministry of Ecology and Environment set nationwide pollution control standards, followed by a 2020document specifying the basic principles, content, calculation methods, and requirements to strongly account for pollutant sources, including the manufacturing of ceramic products.

9 At the same time, large domestic ceramic production enterprises also stepped up their efforts to recycle ceramic waste. Chaozhou's ceramic waste recycling and disposal capacity now meets the needs of the entire city.

10 Oceano, a tile manufacturer with plants in Jingdezhen and Foshan, Guangdong, has implemented a green recycling system. Most production waste is reused internally as raw material for new building products. The Jingdezhen factory, producing 22 million square meters of ceramic materials annually, reuses about 40,000 tons of waste, while the Foshan plant, with 5 million square meters, reuses 10,000 tons.

11 Experts advocate for a "dual-cycle system" to optimize waste management. Large firms should focus on internal recycling, treating and reusing waste to achieve zero emissions from ceramic waste. On a larger scale, industrial parks where ceramic production companies are clustered could establish circular systems to manage waste collectively. Local governments should also introduce more regulations, while the ceramics industry can promote green and low-carbon development. Achieving this goal not only saves energy and reduces emissions, but also helps spread the use of environmental protection

technologies.

An Absorbing Idea

12 Transforming ceramic waste into sustainable building materials is an innovation Yi Design specializes in. Over the past three years, it has developed four types of recycled tiles and bricks. Its laboratory focuses on researching ceramic waste and solid waste as primary raw building materials, with an emphasis on art and aesthetics, to develop sustainable and environmentally friendly materials and products for the construction and design sectors.

13 One standout innovation is its recycled brick, which offers superior water permeability and storage compared with traditional cement bricks. Each brick can hold about 200 milliliters of water, absorbing condensation in cooler conditions and releasing it as temperatures rise, helping to cool urban environment. These bricks also collect rainwater, making them ideal for "sponge cities"—urban areas designed to retain and filter stormwater—or for outdoor paving and gardens.

14 Yi Design's products have been featured in projects like a tiled wall in a Shanghai fashion store and outdoor flooring for a coffee-themed community building. To date, the company has partnered with around 60 factories and over 40 ceramic studios, collecting more than 5,000 tons of waste.

15 Over the past three years, Yi Design's products have won multiple international prizes, including the BLT Built Design Awards Winner 2022 and Red Dot Awards 2022 Winner. In 2023, the company also won the Kering Generation Award.

Practical Uses

16 So far, Yi Design has cooperated with about 60 factories and more than 40 ceramic studios to collect ceramic waste. Over 5,000 tons of ceramic waste has been collected. The company's products have also been used in several projects, including the tiled wall of fashion company COS's store in Shanghai's Century Plaza shopping mall. Their recycled permeable bricks have been used in the outdoor floor area of a coffee-themed community building jointly established by Tongji University and the coffee chain Starbucks in Shanghai.

17 Looking forward, Yi Design aims to reduce costs and explore the possibility of increasing mass production of their materials. While these efforts alone won't eliminate ceramic waste, they raise awareness and provide practical solutions. Many remain unaware of the vast waste piles outside factories, but initiatives like Yi Design's are shifting perspectives, fostering greater environmental respect. More and more people will make their own changes for the environment in the future.

(Adapted from Zhao, 2024)

New Words and Expressions

ceramic /səˈræmɪk/ *n.* 陶瓷

sanitary /ˈsænɪtərɪ/ *adj.* 卫生的

landfill /ˈlændfɪl/ *n.* 垃圾填埋场

kiln /kɪln/ *n.* 窑

glaze /gleɪz/ *n.* 釉

permeable /ˈpɜːmɪəbl/ *adj.* 透水的

condensation /ˌkɒndenˈseɪʃn/ *n.* 冷凝

implementation /ˌɪmplɪmenˈteɪʃ(ə)n/ *n.* 实施

Exercises

Section I Understanding the Text

Task 1 ***Discuss the following questions in small groups.***

1) What environmental risks did traditional ceramic waste disposal methods (e. g., landfilling, repurposing for infrastructure) pose?

2) Compared with traditional cement bricks, what functional advantages do the recycled bricks offer? How might such innovations address broader challenges in densely populated areas?

3) Why is a combination of "market mechanisms" and regulatory frameworks necessary for large-scale industrial recycling?

4) Beyond technological innovation, what societal or cultural changes are needed to reduce ceramic waste generation at its source?

Section II Developing Critical Thinking

More and more Chinese are aware that ecological protection and economic development are interconnected, and safeguarding the environment is essential for maintaining productivity. In this view, improving the ecological environment is not a hindrance but a vital component of long-term economic development. This offers a new path for achieving collaborative coexistence between development and protection—green development.

Task 2 ***Make a presentation on Green Development in China. Your presentation should focus on how China integrates green development into its national strategies, with examples of successful initiatives that demonstrate the positive impact of environmental sustainability on economic productivity and the achievement of sustainable development goals.***

You may prepare your presentation by following the instructions below.

Step 1 Introduce your topic

Explain the concept of Green Development.

Step 2 Structure the main body of your presentation

Choose some examples to illustrate how China is working towards Green Development and the achievements it has made in promoting the harmony between humanity and nature. Below is the structure that you can follow.

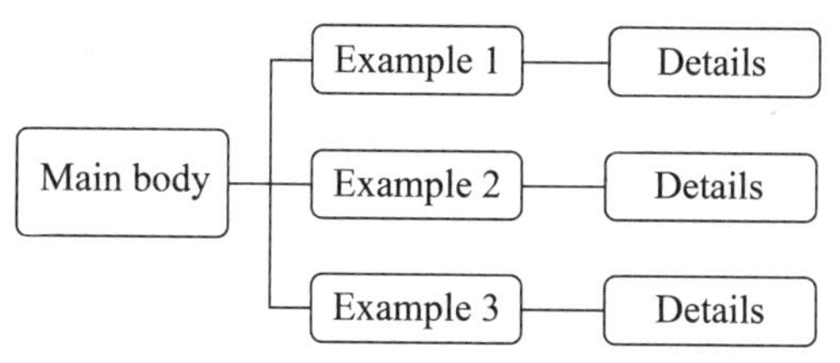

Step 3 End your presentation

In the conclusion of your presentation, you can restate your idea or explain how a college student, who is about to enter society, can contribute to China's Green Development.

UNIT 6

Environmental Problems Since the Twentieth Century

Part One Preparation

Unit Preview

With the continuous acceleration of human industrialization since the 20th century, environmental problems have become increasingly serious, posing severe environmental challenges to human beings.

Currently, the environmental issues that have attracted global attention mainly include global climate change, acid rain pollution, ozone layer depletion, cross-border movement and diffusion of toxic and hazardous chemicals and waste, sharp decline of biodiversity, and marine pollution. Developing counties face widespread ecological and environmental problems, such as water pollution and water scarcity, land degradation, desertification, soil erosion, and deforestation.

With the rapid development of contemporary science and technology, environmental problems caused by high-tech advancements have also become one prevalent. Issues such as nuclear accidents, electromagnetic waves, noise, ozone layer damage caused by supersonic aircraft, and space pollution caused by space flight have attracted significant global attention.

Faced with these environmental issues that threaten human survival, many countries are placing increasing importance on environmental protection. As for China, we believe that respecting, adapting to, and protecting nature is essential for building a modern socialist country in all respects. We must uphold and act on the principle that lucid waters and lush mountains are invaluable assets ensuring harmony between humanity and nature in our development planning.

Learning Objectives

Upon completion of this unit, you will be able to:

- gain a better understanding of the global environmental issues since the 20th century.
- evaluate the environmental damage caused by human activities and developments in high technology.
- write about China's contributions to the biodiversity conservation.

Part Two Global Perspectives

Active Reading 1

Warming Up

Task 1 ***Try to answer the following questions.***

1) Have you heard of the "Dust Bowl" that occurred in America in 1930s to 1940s?

2) What do you know about it and its impact on the American Great Plains?

视听资源

Task 2 ***The dust storms in America from 1930s to 1940s created one of the worst environmental disasters in American history. Watch a video clip and discuss the following questions with your partners.***

1) What happened to New York City, Chicago and Boston in 1934?

2) Where did the dust storms come from?

3) What environmental factors contributed to the dust storms?

Reading

The Dust Bowl in American Great Plains

1 As the majority of the country was dealing with the crippling economic effects of the Great Depression, yet another catastrophe awaited Americans living in the southwestern portion of the Great Plains region—the Dust Bowl. The 1930s and 1940s saw this region devastated by the worst man-made ecological disaster in American history, a series of dust storms that ravaged the land due to a combination of drought and soil erosion.

2 The Great Plains region was settled by thousands of American farmers thanks to the Homestead Act of 1862, which encouraged westward migration by providing settlers with 160 acres of public land. In exchange, these "homesteaders" paid a small fee and were required to live on the land continuously for five years. Most of the farmers raised grazing cattle or grew wheat. Over the years, demand for wheat products grew and consequently millions more acres of prairies grass were plowed and planted for wheat production. At the same time, the introduction of mechanized farming during the Industrial Era had revolutionized the industry. Manual labor was replaced by machinery which could prepare more fields and harvest more crops than ever before.

3 This combination of factors presented a problem when drought struck in 1931. Large dust storms began to sweep across the region. The natural prairie grass could have withstood the

severe drought, but the wheat that was planted in its stead could not. The drought caused the wheat to shrivel and die, exposing the dry, bare earth to the winds. This was the major cause of the dust storms and wind erosion of the 1930s. Dust blew like snow, creating poor visibility and halting road and railway travel. Work crews shoveled the dust from roadways and train tracks, but to no avail. Electric street lights were dimmed by the dark dust, even during the middle of the day. Those motorists who dared to venture out during the storms found that their cars often stalled due to the static electricity the storms created. Small buildings were almost buried. The dust made everyday life miserable. Residents sealed their windows with tape or putty and hung wet sheets in front of their windows to filter out the dust that blew in through cracks in the windows. They covered keyholes, wedged rags underneath doors, and covered furniture with sheets. Everything in the household was covered in a fine layer of dust. Mealtime was especially difficult as cups, plates, dishes, and even food were covered in dust. The dust created health problems for many people; respiratory illnesses were very common. For those living in the Great Plains, life as they had known it had come to a virtual stop. In 1935 homesteader Caroline Henderson wrote to the Secretary of Agriculture Henry A. Wallace to inform him about the grave conditions in which she and thousands of others were living:

There are days when for hours at a time we cannot see the wind-mill fifty feet from the kitchen door. There are days when for briefer periods one cannot distinguish the windows from the solid wall because of the solid blackness of the raging storm...This wind-driven dust, fine as the finest flour, penetrates wherever air can go. After one such storm, I scraped up a panful of this pulverized soil in the first preliminary cleaning of the bathtub! It is a daily task to unload the leaves of the geraniums and other houseplants, borne down by the weight of the dust settled upon them ... A friend writes of attending a dinner where "the guests were given wet towels to spread over their faces so they could breathe". At the little country store of our neighborhood after one of the worst of these storms, the candies in the showcase all looked alike and equally brown ... Dust to eat, and dust to breathe and dust to drink. Dust in the beds and in the flour bin, on dishes and walls and windows, in hair and eyes and ears and teeth and throats. Wind carried the dust hundreds of miles away.

4 By 1937, the dust had reached the Gulf Coast and Middle Atlantic states. The number of dust storms increased from 1934 to 1938. Farmers and local government officials attempted to combat the effects of the storms using soil and water conservation methods like contour lines, a technique which uses terraces and contour planting to minimize water runoff to one end of the field or off the field completely. This technique doubled the odds of a good crop by capturing as much moisture as possible. Despite these efforts, the amount of acreage subject to these storms continued to grow. In her letter to Secretary Wallace, Caroline

Henderson succinctly summed up what, or who, was to blame for their current predicament:

We realize that some farmers have themselves contributed to this reaping of the whirlwind. Under the stimulus of war time prices and the humanizing of agriculture through the use of tractors and improved machinery, large areas of buffalo grass and blue-stem pasture lands were broken out for wheat raising. The reduction in the proportionate areas of permanent grazing grounds has helped to intensify the serious effects of the long drought and violent winds.

5 For the vast majority of those living in the Great Plains, farming the land was their life, their source of sustenance, and their source of income. Without it, they had nothing. Their options were extremely limited. Many people who were unable to make a living on the ravaged land left for places like California where they had heard tales of fertile land and plentiful job opportunities in the agricultural industry. Yet, many stayed on their land. Some aspects of the New Deal, like the Agricultural Adjustment Act, would work to address drought relief. The farmers who stayed hoped that with each passing season, the rains would arrive and the next year's crop yield would be better. The rains eventually did arrive. From 1938 to 1941 the region received a sufficient amount of rain, providing enough moisture to effectively stimulate growth and recovery. The record-breaking rains of 1941 effectively ended the Dust Bowl. The rains coincided with the beginning of World War Ⅱ, and once again agricultural prices began to rise and life began to return to a state of normalcy in the Great Plains.

(Adapted from Smithonian American Art Museum, 2015)

New Words and Expressions

dust bowl /ˈdʌst bəʊl/ *n.* (尤指在美国南部或中部的)风沙侵蚀区,干旱风暴区

catastrophe /kəˈtæstrəfɪ/ *n.* 灾难,灾祸,横祸

devastate /devəˌsteɪt/ *vt.* 彻底破坏,摧毁,毁灭

ravage /ˈrævɪdʒ/ *vt.* 毁坏;损坏;严重损害

acre /ˈeɪkə/ *n.* 英亩

shrivel /ˈʃrɪvəl/ *v.* (使)枯萎,干枯,皱缩

bare /beə/ *adj.* (树木)光秃秃的;(土地)荒芜的

shovel /ˈʃʌvəl/ *vt.* 铲,铲起

stall /stɔːl/ *v.* (使)熄火,抛锚

static electricity /ˈstætɪk ɪˌlekˈtrɪsətɪ/ *n.* 静电

putty /ˈpʌtɪ/ *n.* (主要用来密封窗玻璃)油灰,腻子

wedge /wedʒ/ *vt.* 将……挤入(或塞入、插入)

respiratory /ˈrespərətərɪ/ *adj.* 呼吸的

windmill /ˈwɪndˌmɪl/ *n.* 风车

raging /ˈreɪdʒɪŋ/ *adj.* (自然力)极其强大的,猛烈的

scrape /skreɪp / *vt.* 刮掉,削去

pulverize /ˈpʌlvəˌraɪz/ *vt.* 粉粹,将……磨成粉

preliminary **/prɪˈlɪmɪnərɪ/** *adj.* 预备的，初步的

geranium **/dʒɪˈreɪnɪəm/** *n.* 天竺葵

showcase **/ˈʃəʊˌkeɪs/** *n.* 玻璃陈列柜

contour **/ˈkɒntʊə/** *n.*（地图上表示相同海拔各点的）等高线

contour planting *n.* 等高种植，等高栽植

terrace **/ˈterəs/** *n.* 梯田，阶地

odds **/ɒdz/** *n.*（事物发生的）可能性，概率，机会

acreage **/ˈeɪkərɪdʒ/** *n.* 大块土地；英亩数

succinctly **/səkˈsɪŋktlɪ/** *adv.* 简明地，言简意赅地

predicament **/prɪˈdɪkəmənt/** *n.* 困境，窘境，尴尬的处境

whirlwind **/ˈwɜːlˌwɪnd/** *n.* 旋风，旋流

buffalo **/ˈbʌfəˌləʊ/** *n.* 水牛

bluestem **/ˈbluːstem/** *n.* 须芒草；草原地带的牧草

normalcy **/ˈnɔːməlsɪ/** *n.* 常态

Exercises

Section I Knowledge Focus

Task 1 ***Read the whole passage and complete the following diagram by filling in the blanks with words from the passage.*** （***NO MORE THAN TWO WORDS***）

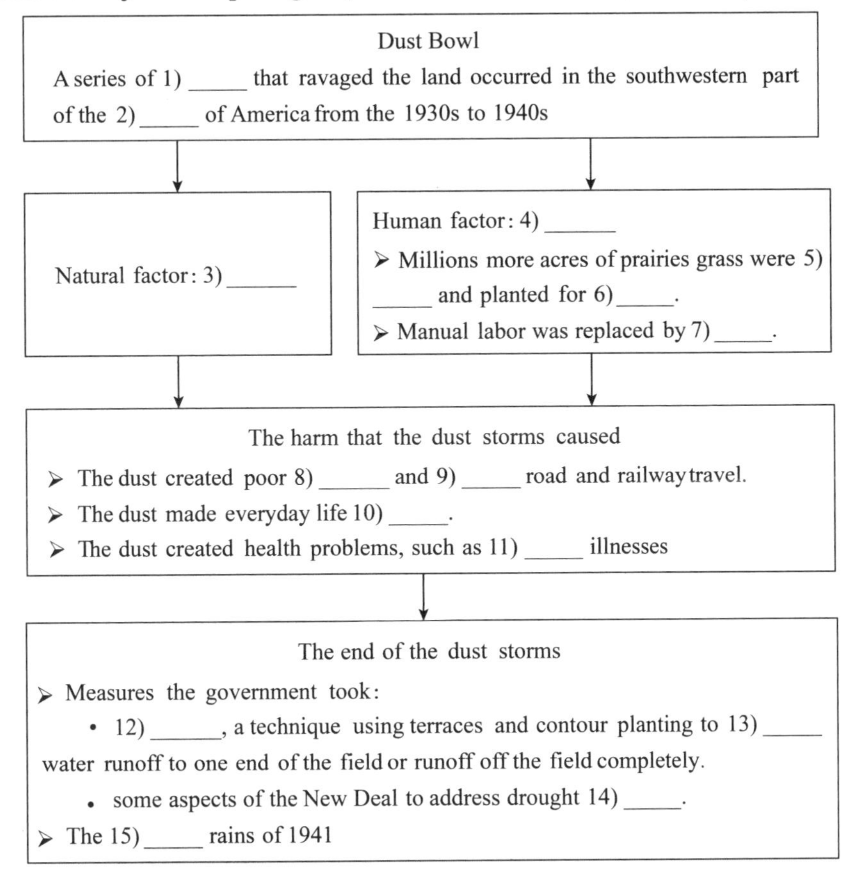

Task 2 ***Given below are five statements. Each statement contains information given in one of the paragraphs of the text. Identify the paragraph from which the information is derived. Answer the questions by writing down the paragraph number (1-5) for each statement.***

__________ A) The farmers in the Great Plains had very few choices except for farming the land, so many of them stayed where they were although that region was devastated by the dust storms.

__________ B) Dust storms caused low visibility, rail and road closures, and almost everything at home was covered with dust. Worst of all, dust could even contribute to respiratory diseases.

__________ C) As America was struggling in the Great Depression from the 1930s to 1940s, the southwestern part of the Great Plains region experienced another disaster.

__________ D) Farmers, to some degree, should be blamed for the disaster because they plowed large areas of grass and pasture lands for wheat production, worsening the adverse effects of long-term drought and wind storms.

__________ E) Unlike the natural prairie grass, which could resist severe drought, the wheat would dry up and die during drought conditions, and the dry and bare land was exposed to the winds, causing dust storms and wind erosion.

Section II Language Focus

Task 3 ***Identify what is being described in the following phrases. Choose your answer from the box below and write the corresponding word in the space provided.***

erosion crack manual terrace sweep

__________ 1) (of work, etc.) involving using the hands or physical strength

__________ 2) (of weather, fire, etc.) to move suddenly and/or with force over an area or in a particular direction

__________ 3) the gradual destruction and removal of rock or soil in a particular area by rivers, the sea, or the weather

__________ 4) a very narrow gap between two things, or between two parts of a thing

__________ 5) one of a series of flat areas of ground that are cut into the side of a hill like steps so that crops can be grown there

Task 4 ***Complete the following sentences with appropriate words or expressions given below. Change the form where necessary.***

respiratory ravage shrivel rage fine

1) The leaves on the plant had __________ up from lack of water.

2) You can use a __________ piece of sandpaper to finish the furniture.

3) The main symptoms of infection with Omicron mutant are similar to those of upper ________ infection, such as fever, dry cough, headache, and sore throat, etc.
4) Forest fires were ________ out of control.
5) Global droughts continue to ________ farmland, intensifying widespread malnutrition and poverty.

coincide combat yield sustenance penetrate

6) The rain had ________ right through his clothes to his skin.
7) The police are now using computers to help ________ crime.
8) You won't get much ________ out of one bar of chocolate.
9) The trees gave a high ________ of fruit this year.
10) It's a pity that your trip to Beijing next week don't ________ with my stay here.

Task 5 ***Find the words in the box that have the same meaning as the underlined words in the following sentences, and write the corresponding word in the space provided.***

serious proportional resist initial disaster

________ 1) Early warnings of rising water levels prevented another major <u>catastrophe</u>.
________ 2) The materials used have to be able to <u>withstand</u> high temperature.
________ 3) The consequences will be very <u>grave</u> if nothing is done.
________ 4) After a few <u>preliminary</u> remarks he announced the winner.
________ 5) The number of accidents is <u>proportionate</u> to the increased volume of traffic.

Task 6 ***Complete the following sentences with the correct form of the words in brackets.***

1) The Homestead Act of 1862 in America encouraged westward ________ (migrate) by providing settlers with 160 acres of public land.
2) Over the years, demand for wheat products grew and ________ (consequence) millions more acres of prairies grass were plowed and planted for wheat production.
3) The introduction of mechanized farming during the Industrial Era had ________ (revolution) the industry.
4) Dust blew like snow, creating poor ________ (visible) and halting road and railway travel.
5) The finding of oil has provided a great ________ (stimulate) to their economy.

Task 7 ***Match the words in the left column with those in the right column to form appropriate expressions. Then complete the following sentences with one of the expressions. Change the form or add articles where necessary.***

seal	effect
halt	the odds
crippling	balance

triple	wounds
ecological	global warming

1) Undoubtedly the high cost of capital has a ___________ on many small firms.
2) The discoveries by researchers show that increased nitrogen levels in seawater upset the natural ___________ of the oceans.
3) In reality, carbon taxes alone won't be enough to ___________, but they would be a useful part of any climate plan.
4) MIT scientists have developed an adhesive that can ___________ or patch a hole caused by a stomach ulcer.
5) Mothers who smoke or drink during pregnancy double or ___________ of their babies becoming violent offenders decades later, even after accounting for other social influences.

Section III Sharing Your Ideas

Task 8 ***The severe dust storms that struck the Great Plains of America during the 1930s and 1940s caused significant disasters for the people. What lessons can we learn from it? What measures has China taken to address the issue of dust storms in our country?***

Active Reading 2

Warming Up

Task 1 ***Have you heard about the Fukushima Daiichi nuclear disaster that happened in Japan in 2011? Do you know how it occurred and what impact it had on the environments?***

视听资源

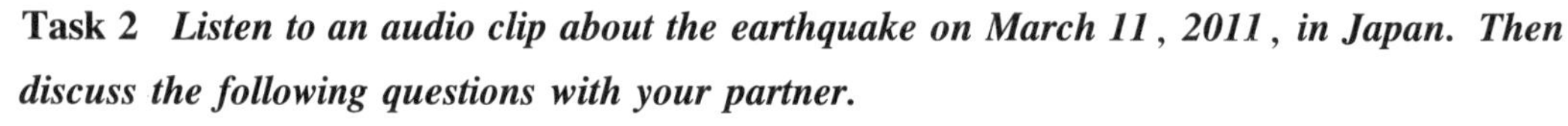
Task 2 ***Listen to an audio clip about the earthquake on March 11, 2011, in Japan. Then discuss the following questions with your partner.***

1) What magnitude was the earthquake recorded by scientists?
2) Where was the quake centered?
3) What damage did the earthquake cause?
4) How far was the quake from Tokyo? And what impact did it have on Tokyo?

Reading

Effects of the Fukushima Nuclear Meltdowns on Environment

1 On the 11th of March, 2011, a magnitude 9.0 earthquake hit Northern Japan. This so-called Tōhoku earthquake led to a tsunami on the Eastern coast of Northern Japan, leading to further destruction. More than 15,000 people died as a direct result of the earthquake and tsunami, more than 500,000 had to be evacuated. The Fukushima Daiichi nuclear power plant was severely damaged by the quake and the tsunami. With no electricity to

power the cooling systems, water inside the reactors began to boil off, causing meltdowns of the uranium fuel rods inside of reactor cores 1 to 3. TEPCO, the company responsible for the plant, began to vent steam from the reactors in order to relieve pressure and prevent a giant explosion. This steam carried radioactive particles out to the Pacific Ocean.

2 Since it became evident that a nuclear meltdown was possibly taking place in the reactor cores, a 20-km zone around the power plant (with an area of about 600 km^2) was declared an evacuation zone and a total of 200,000 people were forced to leave their homes. While evacuations were commencing, multiple explosions destroyed Reactor 1, 2, and 3 and caused a fire of the spent fuel pond of Reactor 4. To cool off the cores, TEPCO took the controversial decision to pump seawater into the reactors. This could not prevent the temperatures from rising even further, as the nuclear fuel rods were left partially uncovered. According to TEPCO, all fuel rods in melted Reactor 1, while 57% melted in Reactor 2 and 63% in Reactor 3. Also, as a result, massive amounts of radioactively contaminated water flowed into the groundwater and back into the ocean.

3 On March 25th, people living in the 30-km radius were asked to voluntarily evacuate their homes and leave the contaminated areas. On April 12, the Fukushima nuclear meltdowns were categorized as a Level 7 nuclear accident—the highest level on the International Nuclear Event Scale (INES), which had previously only been reached by the Chernobyl disaster.

Radioactive Emissions into the Atmosphere

4 The four large explosions, the fire of the spent fuel pond, smoke, evaporation of sea-water used for cooling and deliberate venting of the pressurized reactors all caused the emission of radioactive isotopes into the atmosphere. Measurements of radioactivity taken outside the power plant reached a maximum of 10.85 mSv/h, or about 38,000 times the normal background radiation. Further deliberate venting of block 2 and 3 on March 16 led to additional air-borne releases of radioactivity in similar magnitudes. Radioactivity doses around the plant a week after the earthquake reached levels of up to 1,930 μSv/h—more than 6,000 times normal background radiation.

5 A study by the Norwegian Institute for Air Research (NILU) found that around 16,700 PBq of xenon-133 (250% of the amount released at Chernobyl) were emitted by the Fukushima power plant between March 12th and 19th. This constitutes the largest release of radioactive Xenon in history. Xenon-133 is a radioactive gas with a half-life of 5.2 days, which emits beta-and gamma-radiation and cause harm upon inhalation. Additionally, the NILU study found that 35.8 PBq of caesium-137 (42% of the amount released at Chernobyl) were emitted by the Fukushima power plant between March 12th and 19th. Their study found that radioactive emissions were first measured right after the earthquake and before the tsunami struck the plant, showing that the quake itself had

already caused substantial damage to the reactors. The NILU report also suggests that the fire in the spent fuel pond of reactor 4 may have been the major contributor of airborne emissions, since emissions decreased significantly after the fire had been brought under control.

Soil Contamination

6 The nuclear fallout included different types of radioactive particles, each with its own characteristics. The Japanese Ministry of Science and Technology (MEXT) conducted soil surveys in 100 locations within 80 km of the Fukushima power plant in June and July of 2011. In the entire prefecture, they found contamination with various radioactive substances. While the list of radioactive isotopes released during the meltdowns included more than 30, some are most well-known for causing damage to human tissue, such as Strontium-90.

7 Strontium-90 with a physical half-life of 28 years is a beta-emitting radioactive particle. Upon ingestion, it is metabolized similar to calcium. This means that it is incorporated into the bone, where it can remain for many decades (biological half-life of 50 years). In the bone, strontium irradiates the sensible blood-producing bone-marrow and can cause leukemia and other malignant diseases of the blood. The MEXT survey found strontium-90 in concentrations of 1.8-32 Bq/kg in places outside the 30 km evacuation zone like Nishigou, Motomiya, Ootama or Ono.

Contamination of the Marine Environment

8 Massive amounts of water were used in a desperate attempt to cool the reactors and the burning spent fuel ponds. This led to equally large amounts of radioactive waste water, which was continually discharged into the sea, seeped into soil and ground-water deposits or evaporated into the atmosphere. Between April 4 and 10, TEPCO deliberately released 10,393 tons of radioactive water according to the official report by the Japanese government. Initial estimates of the total contamination of the ocean by TEPCO were 4.7 PBq; however scientists from the Japan Atomic Energy Agency and Kyoto University calculated the total to be 15 PBq, as the amount of radioactive contamination by secondary fallout had been ignored in the initial estimations. Calculations by the IRSN even reached an amount of 27 PBq. Regardless of which calculation is ultimately agreed upon, the Fukushima fallout constitutes the single highest radioactive discharge into the oceans ever recorded. Together with the atmospheric nuclear weapons tests, the fallout from Chernobyl and the radioactive discharge of nuclear reprocessing plants like Sellafield or La Hague, the Fukushima disaster already ranks as one of the prime radioactive pollutants of the world's oceans according to a comprehensive report by the IAEA.

Effects on Food and Drinking Water

9 There is no safe level of radioactivity in food and drinking water. Potentially, even the

slightest amount of radioactivity can cause genetic mutation and cancer. The Fukushima nuclear meltdowns caused a major contamination of food and drink in Japan. According to the IAEA, nearly all vegetable and milk samples collected from Ibaraki and Fukushima prefectures one week after the earthquake revealed levels of iodine-131 and caesium-137 exceeding the radioactivity limits set for food and drink in Japan. In the months after the catastrophe, contamination was found to be even higher in certain foods.

(Adapted from Rosen, 2012)

New Words and Expressions

meltdown /ˈmeltˌdaʊn/ *n.* 核反应堆核心熔毁(导致核辐射泄露)
magnitude /ˈmægnɪˌtjuːd/ *n.* 震级
tsunami /tsʊˈnæmɪ/ *n.* 海啸
evacuate /ɪˈvækjʊˌeɪt/ *vt.* (把人从危险的地方)疏散,转移,撤离
uranium /jʊˈreɪnɪəm/ *n.* 铀(放射性化学元素)
fuel rod / ˈfjuːəl rɒd/ *n.* 燃料棒
reactor core /rɪˈæktə kɔː/ *n.* 反应堆核心,反应堆活性区
radioactive /ˌreɪdɪəʊˈæktɪv/ *adj.* 放射性的,有辐射的
spent /spent/ *adj.* 用过已废的;失效的
contaminate /kənˈtæmɪˌneɪt/ *vt.* 污染,弄脏;毒害
radius /ˈreɪdɪəs/ *n.* 半径范围,周围
pressurize /ˈpreʃəraɪz/ *vt.* 使(潜艇、飞机等内)保持正常气压
isotope /ˈaɪsəˌtəʊp/ *n.* 同位素
measurement /ˈmeʒəmənt/ *n.* 尺寸,长度,数量
background radiation /ˈbækˌgraʊnd ˌreɪdɪˈeɪʃən/ *n.* 背景辐射,本底辐射
airborne /ˈɛəˌbɔːn/ *adj.* 空气中的
xenon /ˈzenɒn/ *n.* 氙;氙气
half-life /ˈhɑːf laɪf/ *n.* (放射性物质的)半衰期
inhalation /ˌɪnhəˈleɪʃən/ *n.* 吸气
caesium /ˈsiːzɪəm/ *n.* 铯
fallout /ˈfɔːlaʊt/ *n.* (核爆炸后的)放射性坠尘
strontium /ˈstrɒntɪəm/ *n.* 锶
ingestion /ɪnˈdʒestʃən/ *n.* 摄取,食入,吸收
metabolize /məˈtæbəlaɪz/ *vt.* 新陈代谢
irradiate /ɪˈreɪdɪˌeɪt/ *vt.* 用射线照射
leukemia /luːˈkiːmɪə/ *n.* 白血病
malignant /məˈlɪgnənt/ *adj.* 恶性的
seep /siːp/ *vi.* 渗,渗透
nuclear reprocessing plant /ˈnjuːklɪə riːˈprəʊsesɪŋ plɑːnt/ *n.* 核后处理工厂
mutation /mjuːˈteɪʃn/ *n.* (生物物种的)变异,突变
iodine /ˈaɪəˌdiːn/ *n.* 碘

Exercises

Section I Knowledge Focus

Task 1 ***Read the first three paragraphs of the passage and complete the following diagram about the damages the Fukushima nuclear meltdowns created by filling in the blanks with words from the passage.*** (*NO MORE THAN TWO WORDS*)

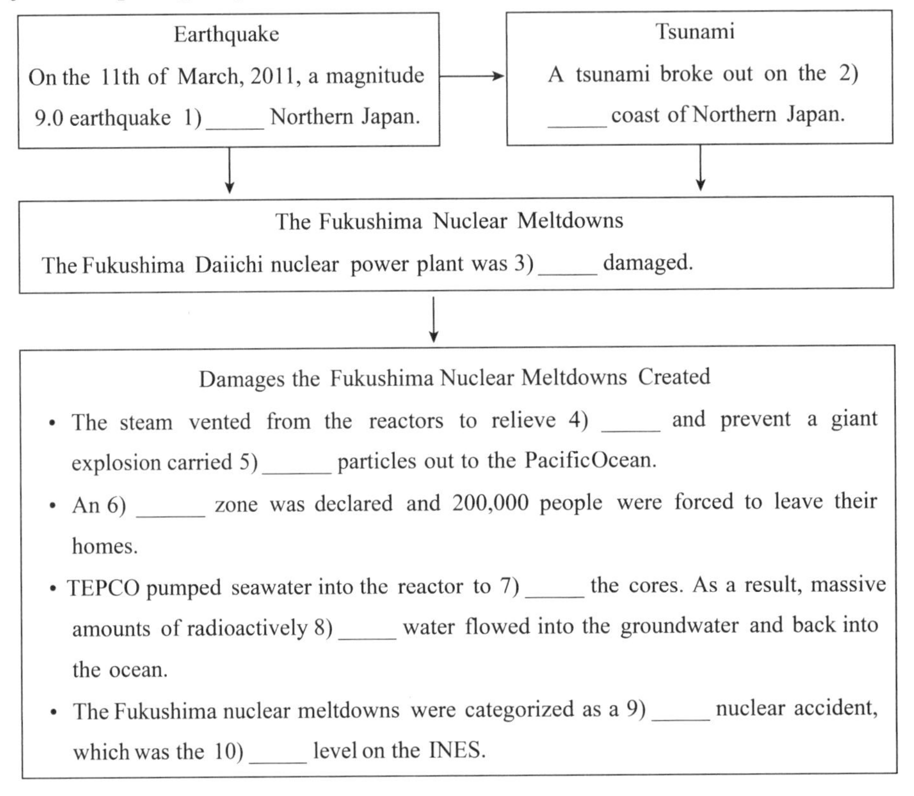

Task 2 ***Read Paras. 4-9 and complete the following table by filling in the blanks with words from these paragraphs.*** (*NO MORE THAN TWO WORDS*)

Different Aspects	Sources for the Contamination	Harms
Atmoshpere	The 1) _____ in the spent fuel pond of reactor 4 was the major contributor of airborne emissions.	Large amount of radioactive gas was emitted, including Xenon-133 and caesium-137.
Soil	The nuclear 2) _____ included various types of radioactive particles.	Strontium-90 irradiates the sensible 3) _____ and can cause leukemia and other 4) _____ diseases of the blood.

(续)

Different Aspects	Sources for the Contamination	Harms
Marine Environment	Massive amounts of water used to cool the reactors and the burning 5) _____ ponds led to equally large amounts of radioactive waste water, which was discharged into the sea, seeped into soil and ground-water deposits or 6) _____ into the atmosphere.	The Fukushima fallout constitutes the 7) _____ highest radioactive discharge into the oceans ever recorded. The Fukushima disaster ranks as one of the prime radioactive 8) _____ of the world's oceans.
Food & Drinking Water	——	Even the slightest amount of radioactivity can cause 9) _____ and cancer. Levels of iodine-131 and caesium-137 in nearly all vegetable and milk samples 10) _____ the radioactivity limits set for food and drink.

Task 3 ***Based on your understanding of the passage, decide whether the following statements are true or false. Write T for true and F for false in the blank provided before each statement.***

__________ A) Facing the meltdowns of the uranium fuel rods of reactor cores 1 to 3, TEPCO decided to release steam from the reactors to pressurize them and prevent a large explosion.

__________ B) Pumping seawater into the reactor was ineffective to keep the temperatures from rising even higher because the nuclear fuel rods were completely covered.

__________ C) Both the Fukushima nuclear meltdowns and the Chernobyl disaster were classified as the highest level of nuclear accident on the INES.

__________ D) The study by NILU found that before the tsunami occurred, the earthquake had already caused considerable damage to the reactors.

__________ E) Even an extremely small amount of radioactivity in food and drinking water is potentially dangerous, so the permissible level of radioactivity in food and drinking water is set very low.

Section II Language Focus

Task 4 ***Identify what is being described in the following phrases. Choose your answer from the box below and write the corresponding word in the space provided.***

massive ultimately dose tsunami secondary

__________ 1) an extremely large wave in the sea caused, for example, by an earthquake

__________ 2) extremely large in size, quantity, or extent

__________ 3) happening as a result of sth else

__________ 4) in the end; finally, after a long and often complicated series of events

__________ 5) the quantity of radiation administered or absorbed

Task 5 ***Complete the following sentences with appropriate words or expressions given below. Change the form where necessary.***

incorporate regardless controversial contributor vent

1) They've now actually had the opportunity to __________ out the anger that they've been having.
2) Sulphur dioxide is a pollutant and a major __________ to acid rain.
3) Universities could __________ IDPs (Individual Development Plans) into their graduate curricula to help students discuss, plan, and achieve their long-term career goals.
4) This money will be paid to everyone __________ of whether they have children or not.
5) Broadway, it turns out, has been a showcase for __________ political themes for generations.

pump deliberate discharge evacuate comprehensive

6) The fish in the river were poisoned by __________ from the chemical factory.
7) His __________ surveys have provided the most explicit statements of how, and on what basis, data are collected.
8) Children were __________ from that village to escape the bombing.
9) This engine is used for __________ water out of the mine.
10) It is not yet clear whether the wiping of records was accidental or __________.

Task 6 ***Match the words in the left column with those in the right column to form appropriate expressions. Then complete the following sentences with one expression. You may need to make some changes.***

relieve	a threat
substantial	need
constitute	impacts
initial	anxiety
desperate	shyness

1) The increase in racial tension has __________ to their society.
2) After she'd overcome her __________, she became very friendly.
3) Psychological intervention can effectively __________ and depression, and improve life quality for cancer patients who underwent surgery and chemotherapy.
4) Technology has caused __________ on our daily life, which can be seen in various aspects.

5) She said there was a __________ for up-to-date and reliable equipment.

Task 7 ***Translate the last paragraph of the passage into Chinese.***

There is no safe level of radioactivity in food and drinking water. Potentially, even the slightest amount of radioactivity can cause genetic mutation and cancer. The Fukushima nuclear meltdowns caused a major contamination of food and drink in Japan. According to the IAEA, nearly all vegetable and milk samples taken in Ibaraki and Fukushima prefectures one week after the earthquake revealed levels of iodine-131 and caesium-137 exceeding the radioactivity limits set for food and drink in Japan. In the months after the catastrophe, contamination was found to be even higher in certain foods.

Section III Sharing Your Ideas

On August 24, 2023, the Japanese government began to discharge nuclear-contaminated water from the Fukushima Daiichi Nuclear Power Plant into the ocean despite strong international opposition.

Task 8 ***Please share your comments on this action based on the following questions.***

1) What are the potential enviromental and economic impacts of this decision?
2) What are the alternative methods Japan could have taken to handle the Fukushima wastewater issue?

Part Three China's Environmental Story

Active Reading 3

Warming Up

Task 1 ***Biodiversity loss has become an issue that should not be ignored. Discuss in groups ways to prevent this trend and strategies to promote its recovery.***

视听资源

Task 2 ***Watch a video clip and answer the following questions.***

1) Why is it important that the total biodiversity of our planet is immense?
2) What benefits do the towering forests that cover 1/3 of the land surface bring to us?
3) Why does the speaker say that our planet biodiversity has dropped dramatically in the last 50 years?

Reading

Biodiversity Conservation in China

1 China is one of the mega-biodiversity countries in the world. But at the same time, it is also one of the countries with the most threatened biodiversity. Biodiversity refers to the ecological complex formed by living organisms (including animals, plants, and micro-organisms), the surrounding environment, and the sum of various ecological processes related to them, including ecosystem diversity, species diversity, and genetic diversity. Biodiversity and sustainable development of human society are interrelated. Biodiversity is the common wealth of all mankind. In June 1992, more than 150 countries signed the Convention on Biological Diversity (CBD) for the common goals of conservation and sustainable use of biodiversity. China was one of the first parties to sign the CBD on June 11, 1992. In December 2016, China was chosen as the host country for the 15th Conference of the Parties of CBD, which was held in Kunming, Yunnan Province, in 2020, which set the new global biodiversity conservation targets for the next decade.

2 The Chinese government and scholars have paid attention to the relationship between biodiversity and human well-being, and have regarded biodiversity as the material basis for sustainable development. The Chinese government has developed programs that promote ecological protection and environmental management under the background of rapid economic growth, which has greatly restored degraded ecological environments, improved provisions of critical ecosystem services, and enhanced rural livelihoods. Chinese scientists

have also made significant contributions to the world in the realm of conservation and the sustainable use of biodiversity. For example, China increased the population of giant pandas (*Ailuropoda melanoleuca*) by establishing nature reserves, and succeeded in artificially breeding them. Yuan Longping, a famous rice breeding expert in China who made outstanding contributions to the world's food security, developed the first strain of hybrid rice in 1970 by crossing the sterile plants of wild rice (*Oryza rufipogon*) found in Hainan with cultivars. Tu Youyou, a Chinese female pharmacist, won the Nobel Prize for her discovery of *Qinghaosu* (*artemisinin*) that helped save millions of lives globally, especially in developing countries. Such achievements in biodiversity conservation have helped lay a solid foundation for the realization of China's 2030 sustainable development goals.

3 Under the concept of "clear waters and lush mountains are invaluable assets", the Chinese government has proposed a series of strategic ideas and goals to build an ecological civilization and a beautiful China. Clear national biodiversity conservation targets have been identified within both the overall national development plan and special ecological plans. These initiatives support China's progress towards an ecologically-sound future based on sustainable growth models and ways of life. After signing the CBD, in June 1994, the National Environmental Protection Agency (currently upgraded to the Ministry of Ecology and Environment, MEE) and other relevant Departments, brought out the "China Biodiversity Conservation Action Plan". In 2010, the former Ministry of Environmental Protection (MEP, currently merged into the MEE), jointly with more than 20 Ministries and Departments, updated the "China National Biodiversity Conservation Strategy and Action Plan (NBSAP, 2011-2030)". The NBSAP identified the strategic goals, strategic tasks, and priority areas and actions for biodiversity conservation in China for the next two decades. And "The National Ecological Function Zoning" released in 2008, along with "The National Main-function Area Plan" released in 2010 have incorporated biodiversity conservation as an important aspect. In 2016, "The 13th Five-year Plan Outline of the People's Republic of China for National Economic and Social Development (2016-2020)" emphasized the need to "strengthen ecological protection and restoration", which requires incorporating major conservation initiatives including the creation of a new National Park system and red line designations for ecosystem conservation. The above-mentioned mainstreaming processes have promoted biodiversity conservation efforts to some degree. However, the contradiction between economic and social development and biodiversity conservation still exists. Sometimes economic development is promoted at the expense of biodiversity when there is conflict. Thus, there is an urgent need to further promote the mainstreaming of biodiversity conservation in decision making and management at all levels of the government, and promote the integration of conservation effectiveness into the assessment system of governmental

officers.

4 In recent years, ecological restoration has been encouraged to aid in the recovery of various degraded ecosystems, such as forest, grassland, wetland, etc. To address the deterioration of forest ecosystems and the reduction in biodiversity, China has been restoring natural forests through six key forest conservation programs, especially the Natural Forest Conservation Program (NFCP) and the Grain to Green Program (GTGP). These restoration programs have generated positive ecological effects by increasing vegetation cover, enhancing carbon sequestration, and controlling soil erosion. Many satellite-derived observations have also verified the corresponding results that most regions of China have experienced a greening trend over the past three decades. For example, the overall forest area in the Loess Plateau grew at an average rate of 600 km^2 per year during the period of the GTGP (from 2007 to 2017). Also, numerous farmers were encouraged to change their income sources from being directly dependent on the land to off-farm jobs, which in turn helped to alleviate the direct dependence on forest resources.

5 The Chinese government has also been implementing other projects to restore degraded grasslands, such as the Beijing-Tianjin Sandstorm Source Grassland Treatment Project since 2002, and the Returning Grazing to Grassland Project since 2003. China is striving to raise the vegetation coverage of grasslands to over 57% by 2025. It is reported that these ecological restoration projects in Inner Mongolia have shown a more positive effect than those in Mongolia. Also, some practical techniques have been developed to restore special degraded ecosystems, especially the alkaline-saline grassland in the Songnen Plain and the degraded black-soil grassland in the Qinghai-Tibetan Plateau. After the massive flooding in the middle and lower reaches of the Yangtze River in 1998, China initiated several large-scale wetland restoration projects to convert reclaimed low-yield croplands back to wetlands. In 2016, the General Office of the State Council issued the "Scheme on Wetlands Protection and Restoration System". The former State Oceanic Administration also issued a guideline for strengthening the management and protection of coastal wetlands in 2017 with the aim of restoring no less than 8,500 hm^2 of coastal wetlands by 2020.

6 These measures have resulted in the significant improvements of the ecological environment and China's wild populations of some threatened species like the crested ibis (*Nipponia nippon*) have seen strong population increases and are reported to be expanding their distribution ranges.

(Adapted from Wang et al., 2020)

New Words and Expressions

biodiversity /ˌbaɪəʊdaɪˈvɜːsɪtɪ/ *n.* 生物多样性
realm /relm/ *n.* 领域；场所
strain /streɪn/ *n.* （动植物的）系，品系，品种
sterile /ˈsteraɪl/ *adj.* 不能生育的，不育的
cultivar /ˈkʌltɪˌvɑː/ *n.* 栽培种，栽培品种
carbon sequestration /ˈkɑːbən ˌsiːkwəˈstreɪʃn/ *n.* 碳储存
alleviate /əˈliːvɪˌeɪt/ *vt.* 减轻，缓和，缓解
alkaline /ˈælkəˌlaɪn/ *adj.* 含碱的；碱性的
saline /ˈseɪlaɪn/ *adj.* 含盐的；咸的
crested ibis /ˈkrestɪd ˈaɪbɪs/ *n.* 朱鹮

Exercises

Section I Understanding the Text

Task 1 ***Discuss the following questions in small groups.***

1) What does biodiversity refer to?
2) What contributions have Chinese scientists made to the conservation and sustainable use of biodiversity?
3) What should we do to avoid developing economy at the expense of biodiversity according to Paragraph 3?
4) How did Chinese ecological restoration programs generate positive ecological effects according to Paragraph 4?
5) What did China do to restore the ecosystems after the flooding in the middle and lower reaches of the Yangtze River in 1998?

Section II Developing Critical Thinking

Task 2 ***Biodiversity is vital for both nature and humans, yet it is dropping fast everywhere in the world. You and your group members are required to conduct research about one endangered species in China, and need to collect relevant information and present a report. Your report should cover the current situation of this endangered species, the reasons contributing to the situation and the measures that China has taken to address the issue. The following steps may help you with the project.***

Step 1 Determine the purpose of your report

Decide whether it is to inform, to warn, or to persuade.

Step 2 Follow the general process of a report

- Generate a list of ideas on the report topic. Create a map by placing the topic in the middle of a page, circling it, and drawing a branch out from it for each related idea, example, or word that describes the topic.
- Collect research from a variety of sources such as Internet, books, articles, interviews

with experts on the topic, speeches or statistical findings. Document the source of the research (title, chapter, page number, database collected from and format of research) so you have the information to properly cite the source.
- Read and analyze the research, highlighting, circling or underlining any important information.
- Draft the report based on the outline you have created.
- Review and edit the report. Pay close attention to organization, logic, grammar, proper citation of sources and whether the report supports the purpose and will answer questions the reader may have on the topic. Revise the report if necessary.
- Proofread the report by identifying any errors in grammar, spelling and mechanics.
- Present your report.

Step 3 Design the structure of your report

A well-structured report generally includes three main parts as shown below.
- **Introduction**: Begin with a clear and concise introduction that provides the audience with the idea or purpose of the report. Use an effective way to hold the audience's attention, such as questions, quotations, statistics, playing the slideshow, etc.
- **Body**: Introduce main points with supporting details, such as examples, statistics or other supporting materials. Begin each paragraph with a main idea, and follow with support from the research collected, citing sources as they are used.
- **Conclusion**: Restate the idea or purpose of the report.

Step 4 Deliver your report

You may refer to the following signposts to make your speech more coherent and impressive.
- **Introducing topic**

  Today, I am going to talk about ...

  The topic of my report is ...

  Today, I will focus on...
- **Presenting a number of main points**

  My first finding is ... /First of all, ... /To begin with, ...

  Moving on to our second finding, ... /Second, ... /In addition, ... /Next, ...

  And finally, our last finding is... /Last but not least, ...
- **Concluding your speech**

  In short, ...

  All in all, ...

  To sum up, ...

  To conclude, ...

UNIT 7

Growth of Environmentalism

Part One　Preparation

Unit Preview

The trajectory of environmental stewardship has transitioned from local grassroots conservation to broad-based engagement in environmental justice and the adoption of innovative governance strategies on a global scale.

American environmentalism had its roots in the 19th century, which initially focused on preserving natural landscapes and resources from the damages of industrialization and urbanization. The modern environmental movement gained momentum in the 1960s, with public awareness rising sharply around the risks of contamination, largely influenced by Rachel Carson's seminal work, *Silent Spring*. Subsequently, environmentalism evolved to include grassroots movements, which emphasized community participation and ecological democracy as integral components.

Today, the environmental movement has matured into a more nuanced understanding of environmental justice. Genuine environmental stewardship can only arise from empowering those most impacted by environmental degradation. For instance, marginalized communities and small farming groups are capable of playing crucial roles in preserving ecological diversity.

China has embarked on an ambitious endeavor to construct a socialist ecological civilization with a focus on green development. Its commitment to pollution control and the promotion of harmonious coexistence between human activities and natural cycles sets a powerful example for global environmental resilience. This approach holds significant potential for contributing to global sustainable development.

These advancements necessitate active engagement and a shift towards a more inclusive strategy to ensure long-term ecological balance.

Learning Objectives

Upon completion of this unit, you will be able to:

- understand the evolution of American environmentalism across four key eras.
- understand the principles and importance of environmental justice globally.
- learn about China's ecological civilization in the context of its development strategy.

Part Two Global Perspectives

Active Reading 1

Warming Up

Task 1 ***Discuss the following questions with your partners.***

1) Have you ever encountered a situation where the use of pesticides was necessary at home? If so, please describe the situation and the specific product you utilized.
2) What impacts do you think they have on both the environment and our health?

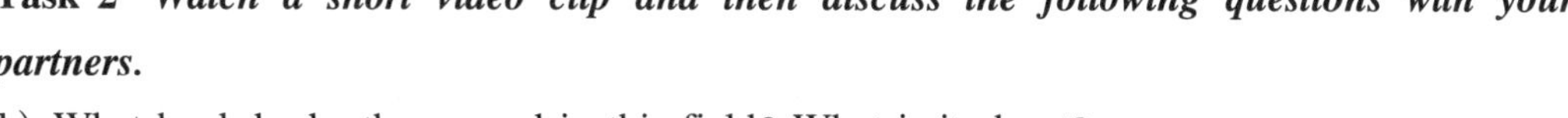

Task 2 ***Watch a short video clip and then discuss the following questions with your partners.***

视听资源

1) What book broke the ground in this field? What is it about?
2) Who is the author of this book?
3) What significant challenges did the author face?
4) What has been the impact of this book?

Reading

Evolution of American Environmental Movement

1 The American environmental movement encompasses a variety of environmental organizations, ideologies, and approaches. The evolution of environmentalism from an ideology into a social movement took place in the following four eras.

The First Era: Conservation and Preservation

2 American environmentalism narratives usually begin with tales of wilderness and the West, whose spectacular landscapes encountered dramatic changes due to urbanization and industrialization. By the 1870s, resource exploitation dominated development patterns in the West. Natural resources were devoured by destructive practices in mining, overgrazing, timber cutting, monocrop planting, and speculation in land and water rights. To protect America's natural resources, environmental organizations arose in support of conservation and preservation.

3 Conservation groups emphasized the efficient use and development of physical resources to combat inefficient land management. Conservationists put forth a developmental strategy based on efficiency, scientific management, centralized control, and organized economic development. This strategy was exemplified by management systems, which were created

to emphasize the balance between immediate and long-term production necessary to sustain a continuous yield.

4 Conservationists established a foothold in American politics in 1901, when President Theodore Roosevelt delineated plans for resource management to Congress. Conservation became the dominant resource strategy of the government during President Roosevelt's tenure, as illustrated by the policies of the new governmental agencies. Moreover, regional and industry-related interest groups emerged as lobbying organizations and agency support groups.

5 During the Roosevelt Administration, however, the first divisions between the conservationists and preservationists emerged. These divisions are best personified by the legendary split between Gifford Pinchot, champion of conservation and efficient land management, and John Muir, cofounder of the Sierra Club. Muir's philosophy advocated natural land management through "right use" of wilderness resources. Preservationists, who believed wilderness preservation to be imperiled by the forces of urbanization and industrialization, viewed traditional conservationist strategies of "right use" and efficient land management as promoting industry needs.

6 Preservation flowed easily from American frontier ideology and notions of the environment as defined in terms of wilderness. According to Muir, it was wrong to view wilderness as simply resources for human consumption; rather, wilderness had an independent value as a "fountain of life". Moreover, the preservationists' vision of nature was romanticized by the poems of William Wordsworth and Henry David Thoreau, which analogized wilderness with religious sacredness, and by Frederick Jackson Turner's classic description of the settling of the frontier and the concomitant rise of a vigorous and independent American democracy.

7 Many environmental organizations that are considered "mainstream" today were formed during the late 1800s, by conservationists and preservationists who desired to protect the natural environment and conserve wildlife.

The Second Era: The Rise of Modern Environmentalism in the 1960s

8 Modern environmentalism differs from the conservation and preservation era in two salient respects. First, whereas the first era emphasized the protection or efficient management of the natural environment, the primary policy of modern environmentalism is based on the cleanup and control of pollution. Second, modern environmentalism displayed "social roots" decidedly absent from the first era. Charted by numerous citizen groups and studies of public attitudes, this change parallels the infusion of particular social values into the public arena and the widespread expression of those values in the environmental arena. Moreover, the approach of modern environmentalism transformed from top-down control

by technical and managerial leaders into bottom-up grassroots demands from citizens and citizen groups.

9 The publication of Rachel Carson's *Silent Spring* in 1962 marked the beginning of modern environmentalism. Carson, known as the "godmother of modern environmentalism", impacted the American public's consciousness with her detailed exposition on the dangers of environmental pollution to human health. By examining the ecological impacts of hazardous substances that pollute both the natural and human environments, like pesticides, Carson fundamentally altered the way Americans perceived the environment and the dangers of toxins to themselves. Emphasizing the problems associated with industrial society, Carson argued that science and technology had been effectively removed from any larger policy framework and insulated from public input and opinion. Carson's controversial thesis not only made *Silent Spring* an epoch event in the history of environmentalism, but also helped to launch a new decade of rebellion and protest in which the concept of "nature" was broadly construed to include quality-of-life issues.

The Third Era: Mainstream Environmentalism

10 Earth Day 1970, widely hailed as the beginning of the third era of American environmentalism, directly resulted from the infusion of social values of the 1960s into environmentalism. Designed to challenge the environmental status quo through peaceful mass mobilization, Earth Day 1970 brought twenty-million Americans together in celebration of quality-of-life issues and concern for the environment.

11 The 1970s also marked the emergence of new issues regarding toxic chemicals, energy, and the possibilities of social, economic, and political decentralization. A seemingly endless series of toxic chemical episodes brought greater publicity, energy, and momentum to the movement. The American citizenry heard about polychlorinated biphenyls in the Hudson River, abandoned chemical dumps at Love Canal and near Louisville, Kentucky, and the Kepone contamination disaster in Virginia.

12 With the establishment of the Environmental Protection Agency (EPA) and the passage of a variety of environmental laws and policies in the 1970s, environmental issues themselves became "mainstream". Primarily comprised of attorneys, engineers, and economists, the EPA developed a complex regulatory structure that categorizes and addresses environmental issues by pollutant and medium.

The Fourth Era: Grassroots Environmentalism

13 Grassroots environmentalism was produced as a reaction to Reagan Administration's anti-environmental policies. In the early 1980s, Ronald Reagan, after acceding to the presidency, implemented many anti-environmental policies by strengthening supervision of regulation and relaxing environmental control, cutting budget and staffers of

environmental protection departments, nominating a conservative to the headship of environmental departments, transferring more environmental responsibility to state and local government and so on.

14 Grassroots environmentalism embraces the principles of ecological democracy, and is distinguished from mainstream environmentalism by its belief in citizen participation in environmental decision making. Perceiving mainstream environmental organizations as too accommodating to both industry and government, grassroots groups utilize "community right-to-know laws, citizen-enforcement provisions in federal and state legislation, and local input in waste methodology and siting decisions". Consequently, although mainstream organizations do perform necessary functions such as educating the middle class to environmental concerns, litigating, and fighting the industrial lobby, grassroots groups create the real movement on issues and force environmentalism onto the public agenda.

15 These new citizen-based groups reflect the evolution of environmentalism from a narrow, wilderness-centered philosophy to a richer, more inclusive ideology encompassing both rural and urban environments. Moreover, grassroots environmentalism cuts across ethnic, racial, and class barriers to introduce a diversity previously absent from the environmental movement.

(Adapted from Stacy, 2004)

New Words and Expressions

encompass /ɪnˈkʌmpəs/ *vt.* 包含,包括

ideology /ˌaɪdɪˈɒlədʒɪ/ *n.* 意识形态,观念体系

monocrop /ˈmɒnəʊˌkrɒp/ *n.* 单一作物

delineate /dɪˈlɪnɪeɪt/ *vt.* 描述,描绘

imperil /ɪmˈperɪl/ *vt.* 危害,使陷入危险

concomitant /kənˈkɒmɪtənt/ *adj.* (尤指相关联的或有因果关系的事)同时发生的,伴随的

salient /ˈseɪlɪənt/ *adj.* 最重要的,显著的,突出的

grassroots /ˈgrɑːsruːts/ *adj.* 基层的,草根的

epoch /ˈiːpɒk/ *n.* 时代,纪元

toxin /ˈtɒksɪn/ *n.* 毒素,毒质

construe /kənˈstruː/ *vt.* 解释,理解

mobilization /ˌməʊbɪlaɪˈzeɪʃn/ *n.* 动员,调动

decentralization /ˌdiːˌsentrəlaɪˈzeɪʃən/ *n.* 分权,去中心化

regulatory /ˈregjulətərɪ/ *adj.* 监管的

supervision /ˌsuːpəˈvɪʒən/ *n.* 监督,管理

conservative /kənˈsɜːvətɪv/ *adj.* 保守的

methodology /ˌmeθəˈdɒlədʒɪ/ *n.* 方法论,方法体系

accede /ækˈsiːd/ *vi.* 就任

lobby /ˈlɒbɪ/ *v.* 游说,斡旋

litigate /ˈlɪtɪgeɪt/ *v.* 诉讼,打官司

Exercises

Section I Knowledge Focus

Task 1 ***How did American environmentalism gradually evolve from an ideology into a social movement? Read the whole passage and complete the following diagram by filling in the blanks with words from the passage. (ONE WORD ONLY)***

The first era	Wilderness exploitation: Early environmental narratives spotlighted the degradation of the West's landscapes due to urbanization and rampant 1) ___________. Rise of conservation movement: It emphasized the 2) ___________ use of resources and gained political ground under President Theodore Roosevelt. 3) ___________ in philosophy: A significant rift developed between conservationists and 4) ___________, who championed the 5) ___________ value of wilderness.
The second era	➢ Advent of modern environmentalism: Initiated by concerns about 6) ___________ and characterized by its roots in broader 7) ___________ issues and citizen involvement. ➢ Impact of *Silent Spring*: Rachel Carson's seminal work heightened awareness of environmental 8) ___________ and reshaped the direction of the movement.
The third era	Earth Day 1970: This event symbolized the 9) ___________ of 1960s social values into environmentalism, leading to public 10) ___________ on a massive scale. 1970s environmental challenges: This decade saw an uptick in environmental crises, notably 11) ___________ contaminations such as the Love Canal incident. EPA's establishment in the 1970s made environmental issues 12) ___________.
The fourth era	➢ Reagan Era's 13) ___________ response: The Reagan administration's 14) ___________ policies in the 1980s led to the rise of grassroots environmentalism, emphasizing local action and 15) ___________ participation.

Task 2 ***Based on your understanding of the passage, decide whether the following statements are true or false. Put T for true and F for false in the blank provided before each statement.***

___________ 1) Despite the transformations in the American environmental movement, its ideologies and principles have remained static throughout the eras.

___________ 2) The conservationist approach to land management was purely motivated by promoting industry needs, with no consideration of long-term ecological sustainability.

___________ 3) Rachel Carson's *Silent Spring* was only a critique of industrial society and had no propositions for larger policy frameworks or public engagement.

__________ 4) Earth Day 1970 was designed as a challenge to environmental status quo by mobilizing peaceful mass protests.

__________ 5) Grassroots environmentalism is limited to addressing environmental issues in urban areas.

Section II Language Focus

Task 3 ***Match the term in the left column with an explanation given in the right column and write the corresponding letter in the space provided below.***

1) dump	A. *n.* the process of creating cities or towns in country areas
2) industrialization	B. *n.* a place where rubbish and waste material are left, for example, on open ground outside a town
3) pesticide	C. *n.* chemicals that farmers put on their crops to kill harmful insects
4) urbanization	D. *n.* the process by which an economy is transformed from primarily agricultural to one based on the manufacturing of goods

1) __________ 2) __________ 3) __________ 4) __________

Task 4 ***Complete the following sentences with appropriate words or expressions given below. Change the form where necessary.***

concomitant conservative decentralization delineate exploitation

1) __________ of environmental policies allowed local communities to tailor solutions to their unique environmental challenges.

2) The environmental report clearly __________ the areas most vulnerable to deforestation.

3) The rapid industrial growth in the city had the __________ effect of increased air pollution and deteriorating health conditions among residents.

4) The unchecked __________ of natural resources threatens the delicate balance of our ecosystems.

5) __________ policymakers often prioritize economic growth over environmental protection, viewing stringent regulations as a hindrance to development.

litigate lobby mobilization regulatory supervision

6) Many businesses find it challenging to navigate the complex __________ landscape related to environmental standards.

7) Environmental groups decided to __________ against the company responsible for the oil spill, seeking compensation for the damage done to the coastline.

8) Several green organizations teamed up to __________ for stricter regulations on single-use plastics.

9) Proper __________ of industrial activities is essential to ensure that companies adhere to environmental safety standards.

10) The global __________ against climate change saw millions take to the streets, demanding urgent action from their governments.

Task 5 ***Find the words in the box that have the same meaning as the underlined words in the following sentences, and write the corresponding word in the space provided.***

construe encompass imperil mobilization salient

__________ 1) The new environmental protection acts include a wide range of measures, from air quality standards to wildlife conservation.

__________ 2) Different activists might interpret the government's environmental policies in varied ways, some seeing them as progressive, while others believe they don't go far enough.

__________ 3) Polluting rivers and streams can harm both aquatic life and the communities that depend on these water sources.

__________ 4) One of the most noticeable issues in the environmental movement is the urgent need to reduce carbon emissions.

__________ 5) The local environmental group called for arms with over 1,000 volunteers who, in 1993, heeded the call for a citywide cleanup and tree-planting campaign to avert urban pollution.

Task 6 ***Match the words in the left column with those in the right column to form appropriate expressions. Then complete the following sentences with one of the expressions. Change the form or add articles where necessary.***

delineate	alarm bells
implement	policies
impact	consciousness
mark	beginning
sound	plans

1) The city council __________ to increase green spaces, making urban areas more environmentally friendly.

2) The environmental agency __________ for restoring the wetlands, aiming to bring back native species.

3) The sudden decline in bee populations across the world __________ among environmentalists, prompting urgent research into the phenomenon.

4) The global agreement on reducing plastic waste in oceans __________ of a united effort against marine pollution.

5) The groundbreaking documentary on climate change set out to __________ of viewers worldwide to make them aware of the strong need to take environmental action.

Section III Sharing Your Ideas

Task 7 ***Early environmental movements focused on the management and protection of natural resources, while modern environmentalism emphasizes pollution control, grassroots demands, and inclusivity of diverse populations. How have turning points affected the direction of environmentalism?***

Active Reading 2

Warming Up

视听资源

Task 1 ***Think about your day so far. How many times did you come across single-use plastics? Are there viable sustainable alternatives that you might consider using instead?***

Task 2 ***Watch a Ted talk and discuss the following questions with your partners.***

1) What is the main topic of this speech?

2) What negative consequences of plastic does the speaker mention?

3) What is "Cancer Alley" and why is it called so?

4) How does the speaker connect the idea of disposability with social justice?

5) What is biomimicry?

Reading

Biodiversity Meets Environmental Justice

1 In 1990, I was just starting as a young assistant professor at the University of Michigan's School of Natural Resources and Environment. That same year, inspired by the publication of *Toxic Wastes and Race* three years earlier, Dr. Bunyan Bryant and Dr. Paul Mohai organized the "Michigan Conference on Race and the Incidence of Environmental Hazards". Bunyan and Paul invited me to participate in the conference and write a chapter on the impacts of pesticides on farm workers and their international dimensions. Being an agroecologist and a conservation biologist trained as an ecologist, the topic of the chapter fell somewhat outside my area of expertise. However, I took on the challenge and for the next six months immersed myself in the topic of pesticides and environmental justice. *Toxic Wastes and Race* and the Michigan Conference had an immense impact on my future career as a scholar. From the conference I emerged energized and decided to examine the connections between the discipline of ecology and environmental justice. Almost twenty years have passed, and this year I participated with my partner John Vandermeer in the first Environmental Justice Symposium at the Annual Meeting of the Ecological Society of America (ESA).

2 Among the myriad political issues that are of concern to ESA, three stand out as not only

important in their own right, but together take on a particular urgency: environmental justice, globalization and tropical conservation. The environmental justice movement has focused on the urgent contemporary task of documenting and struggling against political and economic decisions that place underprivileged groups at environmental risk. For example, Memphis, site of the 2007 ESA meetings, has communities of mainly African Americans who remain subject to the environmental hazards that originally stemmed from the production of chlordane by the Velsicol Company (the same chemical and same company, by the way, that was so active in attempting to block the publication of *Silent Spring*). Even though the use of chlordane was banned in the U. S. in 1988, residues remain in soil and sediments throughout the Memphis area, especially in areas populated by low-income families and people of color. This was one of the topics of my paper at the Michigan Conference.

3 Chlordane itself provides a bridge to the next political issue we identify as critical, the political debate associated with recent trends of globalization. When chlordane was banned for use in the U. S., as so frequently happens, Velsicol simply changed marketing strategies and began shipping its now acknowledged dangerous chemical to the unwitting farmers of the Developing World. The globalized economy certainly aided Velsicol at a time when it faced a clear under-consumption crisis (no market for a product it was geared up to produce in large quantities). The small farmers and farm workers in Latin America, Africa and Asia thus became victims of an environmental injustice that had a clear ecological connection to the African American community in Memphis. In most recent times those small farmers and farm workers of the Global South have not been sitting idly by as the contemporary globalization trend sends a tide that threatens them, but they have been major participants in one of the largest grassroots movements in the history of the world, the movement commonly referred to as the "anti-globalization" movement.

4 Those small farmers sit in the midst of what is, for non-human nature, one of the most important places in the world—the agroecosystems that surround the remaining patches of natural habitat in the vast majority of the world's tropical terrestrial ecosystems. What we now know about the functioning of tropical ecosystems convinces us that the environmental injustice faced by these small farmers and farm workers, so similar in its political overtones to that faced by minority communities throughout the Developed World, has an inevitable connection to the political issue that probably inspires members of the ESA more than any other, that of the conservation of tropical biodiversity.

5 These three political movements are intrinsically interconnected and should not be viewed in isolation. Our argument is founded not on a bed of political thought, but rather emerges from what contemporary ecology tells us about the organization of biodiversity.

6 Conservationists in the past have focused on the purchase and protection of large tracts of

land. From what we now know about how biodiversity is structured ecologically, this is a doomed strategy. While there is no rational need to convert any more forests to agriculture, and we join in with others who seek to preserve whatever remaining natural habitat exists in the world, they are in fact being converted and the future almost certainly will present us with mainly fragmented landscapes. It is in those fragmented landscapes that the world's biodiversity will be located. A long-term plan for biodiversity conservation needs to acknowledge that fact and work at the landscape level to not only preserve the patches of native vegetation that remain, but also to construct a landscape that is "migration-friendly". That landscape is most likely to emerge from the application of agroecological principles. Those principles are most likely to be enacted by small farmers with land titles, who are a consequence of grassroots social movements. Indeed, it would be only slight exaggeration to suggest that these social movements in fact hold the key to real biodiversity conservation.

7 If we allow ourselves to be constrained to the ever-shrinking area of formally protected areas, we accede to the enemies of biodiversity conservation the millions of fragments of natural habitat that today probably contain most of the world's biodiversity. Joining the struggle of the millions of small farmers all over the world is as much part of the environmental justice movement as joining the struggle of African Americans in Memphis for a cleaner environment. Seeing the connections between these struggles is a sign of the maturity of the Environmental Justice Movement.

(Adapted from Bullard et al. ,2007)

New Words and Expressions

agroecologist /ˌæɡrəʊˌiːˈkɔlədʒɪst/ *n.* 农业生态学家

ecologist /ɪˈkɔlədʒɪst/ *n.* 生态学家

justice /ˈdʒʌstɪs/ *n.* 正义

myriad /mɪrɪvd/ *n.* 无数，大量

underprivileged /ˌʌndəˈprɪvɪlɪdʒd/ *adj.* 贫困的

chlordane /ˈklɔːdeɪn/ *n.* 氯丹

residue /ˈrezɪdjuː/ *n.* 残留物

unwitting /ʌnˈwɪtɪŋ/ *adj.* 不知情的，无意的

under-consumption /ˌʌndəkənˈsʌmpʃən/ *n.* 消费不足，消耗不足

anti-globalization /ˌæntɪˌɡləʊbəlaɪˈzeɪʃən/ *n.* 反全球化

agroecosystem/ æɡrəʊˈiːkəʊsɪstəm/ *n.* 农业生态系统

terrestrial /təˈrestrɪəl/ *adj.* 陆地的

intrinsically /ɪnˈtrɪnsɪklɪ/ *adv.* 本质地，内在地

agroecological /ˈæɡrəʊˌiːkəˈlɔdʒɪkəl/ *adj.* 农业生态(学)的

accede /əkˈsiːd/ *vi.* 同意

Exercises

Section I Knowledge Focus

Task 1 ***Read the whole passage and complete the following diagram by filling in the blanks with words from the passage. (ONE WORD ONLY)***

Introduction (Para. 1)	➢ Involvement in the Michigan Conference and writing on the impact of 1)__________ signify a deep dive into environmental justice.
Body (Paras. 2-6)	➢ Key issues: Environmental justice, globalization, and tropical conservation are 2)__________ and 3)__________. ➢ Environmental justice: The example of Memphis highlights how the environmental 4)__________ from chlordane production disproportionately impacted 5)__________ African American communities. ➢ Globalization impact: · After being 6)__________ in the U. S., the sales of chlordane shifted to the 7)__________ World countries. · Small farmers in Latin America, Africa, and Asia became 8)__________ of environmental injustice due to globalization. ➢ Biodiversity conservation: · The essential role of 9)__________ and 10)__________ farmers in conserving tropical biodiversity. · The link between 11)__________ movements and effective biodiversity conservation.
Conclusion (Para. 7)	➢ The Environmental Justice Movement must extend beyond 12)__________ areas, linking small farmers, and African Americans, struggles to prove its 13)__________.

Task 2 ***Based on your understanding of the passage, decide whether the following statements are true or false. Put T for true and F for false in the blank provided before each statement.***

__________ 1) The Environmental Justice Movement's maturity is underscored by its recognition of the interconnectedness between environmental justice and biodiversity conservation efforts.

__________ 2) The "Michigan Conference on Race and the Incidence of Environmental Hazards" was limited to discussing the ecological impacts of pesticides without addressing racial disparities.

__________ 3) The article credits the ban of chlordane in the U. S. as a significant victory for environmental justice, without acknowledging its continued impact in other regions.

__________ 4) The author argues that globalization has uniformly benefited small farmers in the Global South by opening new markets for their products.

__________ 5) The passage suggests that the Environmental Justice Movement has fully integrated biodiversity conservation into its agenda from its inception.

__________ 6) The conclusion of the article proposes that the key to real biodiversity conservation lies exclusively in the enforcement of stricter global environmental laws.

Section II Language Focus

Task 3 ***Identify what is being described in the following phrases. Choose your answer from the box below and write the corresponding word in the space provided.***

agroecology	agroecosystem	biodiversity	globalization	sustainability

__________ 1) variety of life in ecosystems, species, and genetic levels.

__________ 2) farms considered as interconnected natural systems

__________ 3) ensuring resources last for future generations' use

__________ 4) applying natural processes to farming for smarter farming practices

__________ 5) worldwide integration of economies, cultures, and governments

Task 4 ***Complete the following sentences with appropriate words or expressions given below. Change the form necessary.***

accede	globalization	intrinsically	justice	migration

1) Biodiversity is __________ valuable, providing essential ecosystem services that support life on Earth, including pollination, water filtration, and climate regulation.

2) Animal __________ patterns have been profoundly affected by climate change, with species moving to new areas in search of suitable habitats and resources.

3) In recent environmental negotiations, several nations have __________ to international agreements aimed at reducing carbon emissions and protecting biodiversity, signaling a commitment to global environmental stewardship.

4) In the late 20th century, the environmental __________ movement emerged, advocating for the quality of being fair and reasonable by addressing the disproportionate impact of environmental hazards on marginalized communities.

5) The __________ of the 21st century has accelerated environmental change, spreading the benefits and burdens of economic development unevenly across the globe.

myriad	sediment	unwitting	under-consumption	underprivileged

6) The history of conservation efforts showcases a __________ of strategies, from the establishment of national parks to community-based conservation, reflecting diverse approaches to preserving natural habitats.

7) Environmental history often highlights the plight of __________ communities who suffer the most from industrial pollution, yet have the least resources to combat its effects.

8) Studying the ___________ layers within a lake can reveal centuries of environmental change, including the impact of human activities on natural ecosystems.

9) Many consumers are ___________ participants in environmental degradation, unaware that their choices contribute to deforestation and pollution on a global scale.

10) During the Great Depression, ___________ became a significant issue, as decreased demand for goods led to increased unemployment and unused natural resources.

Task 5 ***Translate the following Chinese expressions into English with what you have learned from the text.***

1) 具有特别的紧迫性　2) 环境不公的受害者　3) 带来了一个浪潮
4) 应用农业生态学原则　5) 由小农执行　6) 碎片化的景观
7) 内在相互连接的　8) 基层运动　9) 生物多样性保护
10) 环境正义运动的成熟

Task 6 ***Translate the following paragraph from Chinese into English, using the phrases and expressions from Task 5.***

在环境挑战日益严峻的当下，生物多样性保护刻不容缓。这不仅源于自然保护的客观需要，也出于对遭受环境不公待遇群体的关注。小规模农户积极响应生态农业原则，在构建可持续及有韧性的农业实践中，农业生态原则发挥着核心作用。这些努力是全球争取更公平、更环保的基层运动的一部分。随着这些运动的不断深入，变革浪潮在碎片化景观中涌起，凸显环境保护与社会公正之间的内在联系。种种举措之下，环境正义运动迈向成熟，标志着我们在理解和应对地球挑战方面，迈出了重要一步。

Section III　Sharing Your Ideas

Task 7 ***It is estimated that more than 3 million tons of pesticides are used annually in the world. Watch a short video weighing its pros and cons. Then fill out the following chart after discussing with your partners on the topic of the impact of pesticides, drawing on both Rachel Carson's pioneering insights highlighted in Active Reading 1, the real-world implications analyzed in Active Reading 2, and this video.***

视听资源

Impact of Pesticides: Rachel Carson's Work & Global Implications

Introduction	Rachel Carson, in her groundbreaking work *Silent Spring* (1962), was among the first to raise concerns about the extensive use of pesticides, particularly DDT, and its potential long-term effects on the environment and human health. Her work acted as a catalyst for environmental activism and led to significant changes in pesticide regulation.

（续）

Impact	Environmental impact	
	Impact on communities	
	Broader global implications	
The road ahead	Alternatives to pesticides	
	Policy changes	
Conclusion		

Part Three China's Environmental Story

Active Reading 3

Warming Up

Task ***Chinese have pushed back vast stretches of deserts and turning them into thriving forests. Discuss with your partners how this is achieved.***

Reading

China Makes New Successes in Desertification Prevention and Control

1 The Taklimakan, the largest desert in China, has been completely encircled by a green belt stretching 3,046 km as of late November 2024, thanks to more than four decades of efforts as part of China's Three-North Shelterbelt Forest Program (TSFP), the world's largest afforestation program to tackle desertification.

2 This great achievement has captured worldwide attention. The international community observes that the TSFP has not only contributed to the global combat against desertification, but also increased the global forest coverage rate. They praised the program as the "green Great Wall" that China fortifies to fend off the challenges arising from climate change. The Taklimakan Desert is China's largest and the world's second-largest drifting desert. The completion of the ecological barrier vividly demonstrates China's battle against desertification.

3 After over 40 years of unremitting efforts, China has blazed a special path of desertification prevention and control with Chinese characteristics. Environmental conservation and improving people's well-being have entered a virtuous cycle. The country has become a global model for desertification control. So far, China has effectively rehabilitated 53 percent of its treatable desertified land, with the area of desertified land decreasing by 65 million mu (about 4.3 million hectares). It is the first country in the world to achieve zero growth in land degradation, and the first to reduce the area of desertified and sandy land.

4 To be specific, Saihanba in North China's Hebei province has transformed from a desolate land into the world's largest man-made forest, thanks to the dedication of three generations of foresters to the Saihanba afforestation project; most of the sand land in the Mu Us Desert, which stretches from North China's Inner Mongolia Autonomous Region to

Shaanxi province, has been brought under control; the Kubuqi Desert, China's seventh-largest desert, has explored a brand new approach to combating desertification through industrial development.

5 The United Nations Convention to Combat Desertification (UNCCD) Secretariat has twice honored China for its "outstanding contribution to combating desertification", praising its significant contributions to global desertification control. The 16th Conference of the Parties (COP16) to the UNCCD held in Riyadh, capital of Saudi Arabia, set up the China Pavilion, which marked the first time that China showcased its achievements in desertification control, particularly through the TSFP. "I was moved by the images of different generations of Chinese people fighting desertification and by China's leadership in this process", said Andrea Meza Murillo, UNCCD deputy executive secretary. She emphasized that effective policies, community and local government involvement, as well as innovation and technology, are key components of China's success.

6 The achievements in desertification control reflect China's relentless efforts to advance ecological conservation. Since the 18th National Congress of the Communist Party of China, the country has ensured stronger ecological conservation and environmental protection across the board, in all regions, and at all times. China has achieved a significant transformation from remediation of major areas to systematic governance; realized an important shift from passively responding to ecological issues to taking proactive action to address them; has become a leader from a participant in global environmental governance; and has realized a major shift from practical-exploration-based conservation to the one with theoretical guidance.

7 China's forest coverage ratio and forest stock volume have both been on the rise for 40 consecutive years. The country ranks first in the world in terms of forest resource growth and afforestation area. It has contributed to one fourth of the world's newly added green areas. China is also contributing to the global fight against desertification with firm and concrete actions.

8 The year 2024 marks the 30th anniversary of China's adoption of the UNCCD. Over the past three decades, China has been actively involved in global desertification governance within the framework of UNCCD. It promoted the establishment of the Committee for the Review of the Implementation of the Convention (CRIC), setting up the strategic framework and objectives; established an office for UNCCD implementation and formulated a national action plan; and set up regional mechanisms to promote regional cooperation on UNCCD fulfillment. Besides, China hosted the COP 13 to the UNCCD and carried out cooperation on desertification control within the framework of the Belt and Road Initiative (BRI). It has held nine sessions of the Kubuqi International Desert Forum to promote international policy dialogue and exchanges.

9 China is also actively sharing its technologies and experiences of desertification control with other developing countries, pursuing green development together with the Global South. It has established an international training center and an international knowledge management center on combating desertification together with the UNCCD Secretariat. Besides, the country trains nearly 100 professionals in this field each year for developing countries in Africa, Asia and Latin America, by hosting international seminars and setting up demonstration bases for desertification control. It has taken the initiative to align with the Great Green Wall initiative launched by the African Union and the Saudi Arabia-led Middle East Green Initiative, within the frameworks of the Forum on China-Africa Cooperation (FOCAC) and the China-Arab States Cooperation Forum (CASCF). It has also established the China-Arab International Research Center on Drought, Desertification, and Land Degradation.

10 The prevention and control of desertification is a great cause bearing on the sustainable development of all humanity. China will keep acting as a participant and leader in global desertification control, and work with all parties to advance ecological conservation and promote sustainable development for a clean and beautiful world.

(Adapted from Wang, 2024)

New Words and Expressions

desertification /dɪˌzɜːtɪfɪˈkeɪʃən/ *n.* 荒漠化

afforestation /əˌfɒrɪˈsteɪʃən/ *n.* 植树造林

unremitting /ˌʌnrɪˈmɪtɪŋ/ *adj.* 不停的

rehabilitate /ˌriːəˈbɪlɪteɪt/ *vt.* 修复

remediation /rɪˌmɪdɪˈeɪʃən/ *n.* 补救

Exercises

Section I Understanding the Text

Task 1 ***Discuss the following questions in small groups.***

1) What key factors must other countries consider to replicate China's afforestation success?

2) Why is China's shift from "practical-exploration-based conservation" to "theoretical guidance" significant for environmental governance?

3) What are the environmental and economic consequences of China's large-scale afforestation projects?

4) How does China's leadership in global environmental governance align with its broader geopolitical and diplomatic strategies?

5) How can future technological advancements further enhance China's desertification control efforts?

Section II Developing Critical Thinking

In China, environmental protection is increasingly seen as integral to the country's development, aligning with the socialist values of harmony between human society and nature. The emphasis on sustainability reflects a broader understanding that ecological conservation is essential for long-term prosperity.

Task 2 ***The English Club at your university is holding a writing competition to enhance students' environmental consciousness. The participants should research, synthesize, and communicate a local environmental issue in a concise manner.***

You may prepare your writing by following the instructions below.

Step 1 Research and choose an issue

- **Method:**

  Walk around your local community or campus to observe environmental conditions firsthand.

  Conduct short interviews with community members to gather their perspectives on local environmental concerns.

  Browse local newspapers, blogs, or online community forums for information on environmental issues.

- **Tips:**

  Choose a topic that has a visual aspect so it can be effectively illustrated or photographed.

  Prioritize issues that have actionable solutions, even small ones.

Step 2 Draft your report

- **Title:**

  Purpose: To immediately catch the reader's attention.

  Tips: Keep it short. Use powerful adjectives.

- **Description:**

  Purpose: To give a brief overview of the problem.

  Tips: Start with a general statement, then specify the local issue.

- **Suggested action /solution:**

  Purpose: To offer a tangible solution to the problem.

  Tips: Think of what local residents or students can realistically implement.

- **Illustration or photo:**

  Purpose: To visually communicate the essence of the issue.

  Tips: Ensure the visual complements your written content.

- **Fact or statistic:**

  Purpose: To provide evidence or context to your report.

  Tips: Make sure your fact is recent and relevant. Always provide a source for your statistic.

Step 3 Format & present your report

- **Method:**

  Use word processing software or design apps for a polished look.

  Ensure that the text is easily readable.

- **Tips:**

  Use bullet points or numbered lists to break down complex information.

  Use a consistent color scheme and font style throughout the report.

Step 4 Class sharing (Optional)

- **Method:**

  Prepare a short 2-minute presentation based on your report.

  Practice your pronunciation and intonation to effectively communicate your issue and solution.

- **Tips**

  Use visual aids (e.g., slides, photos) to support your presentation.

UNIT 8

Towards a Sustainable Future

Part One Preparation

Unit Preview

With the ever-increasing scale of environmental problems, it is imperative to foster ecological sustainability through innovative and strategic approaches.

A key step in this direction is restoring the power of natural ecosystems as the vital basis for ecological health. Rewilding large land and sea areas can prevent the loss of biodiversity while strengthening the resilience of the biosphere.

In addition to restoration, agricultural innovations, such as carbon farming, play a vital role. By increasing soil carbon sequestration, agriculture can shift from a carbon source to a carbon sink with financial incentives and supportive policies.

Building on these efforts, China has embarked on its ambitious journey toward a low-carbon economy. Its commitment to achieving carbon neutrality by 2060 demonstrates how environmental goals can align with economic growth, technological innovation, and energy security.

Taken together, restoration, innovation, and effective policy form the foundation of a harmonious and sustainable coexistence with the planet.

Learning Objectives

Upon completion of this unit, you will be able to:

- understand the concepts of Earth's abundance, carbon farming and low-carbon transition
- analyze claims about wilderness protection and carbon farming
- evaluate China's low-carbon transition and its economic and environmental impacts

Part Two Global Perspectives

Active Reading 1

Warming Up

Task 1 ***Discuss the following questions with your partners.***

1) How do you feel when surrounded by a lush forest compared to standing in a deforested area?
2) What do these two contrasting landscapes indicate about our environment?

Task 2 ***Watch a short video clip. Then answer the following questions.***

1) According to the video, what is mass extinction?
2) How many mass extinctions have occurred throughout Earth's history, and what are some potential factors that have contributed to these events?
3) What do current extinctions tell us about our impact on modern environment?

Reading

Restoring Abundant Earth

1 Moving toward an ecological civilization begins by embracing the epic, deep truth of Earth as a living planet that creates diversity, complexity, biological wealth, and a stunning array of forms of awareness. Wilderness—or free nature by any other word—is the cauldron within which Earth performs the alchemies that create such splendor. Recreating civilization will have us turning toward designing a thriving coexistence of human and nonhuman communities, interpenetrating human history with natural history into uncharted realms of beauty, diversity, plenitude, and freedom for all.

2 The ethical ramifications for nonhumans and humans are ultimately not discrete: they intersect in the profound space we call dignity. Restoring big wilderness in the biosphere restores dignity to all beings to be and live as who they are—not as starving, persecuted, and exiled refugees teetering on an edge between living and dying, or existence and extinction. It also restores dignity to human beings, for the rampage called "civilized behavior" on Earth promotes and institutionalizes such rank attributes as greed, arbitrary exercise of power, and arrogance, and incites killing, experimenting on, harming, and enslaving beings with impunity.

3 The most encompassing context for the thriving of all life is sustaining the biosphere's

freedom to express its inherent nature. The biosphere's nature is to create diversity of life-forms, plenitudes of wild beings, complex and dynamic ecologies, and extraordinary living phenomena such as animal migrations, biodiverse ancient forests, intricate mycorrhizal networks, and fascinating variations of intelligence and awareness in all life's kingdoms. For the biosphere to be free to express its nature, we must pledge vast areas of continents and ocean to remain unoccupied, unexploited, and connected. Being wilderness is the original blueprint of the biosphere and the precondition to express itself "as a work of art that gives birth to itself".

4 For rewilding the biosphere and ourselves within it, I propose three frameworks to think with that provide pragmatic and visionary directions forward. The first and most immediately achievable goal is robustly protecting 25-75 percent—or a rough and memorable half—of Earth's biomes. This is known as the platform of "Nature Needs (at least) Half" or in E. O. Wilson's wording, and recent book title, *Half Earth.*

5 Protected areas today are the havens for Earth's remaining biological wealth, ecological complexity, and evolutionary potential. They must remain formally protected, until the time comes when such areas are no longer needed. The entire Earth can then become what David Brower envisioned as "a conservation district in the universe"—Earth Park, except that the word "park" will be as unnecessary as human-nature legal boundaries. Protected areas, however, are indispensable until that day when human beings create a way of life that has left in the dustbin of history all commerce in wildlife body parts, as well as such irrational trade-offs as wetlands for cane sugar; rain forests for meat and palm oil; prairies for corn and soy; intact ecosystems for coltan, diamonds, gold, or oil; mountains for coal; sagebrush landscapes for natural gas; boreal forest for butamine; and a life-filled ocean for trash dumping and mass-extermination-begotten seafood. Until an enlightened time dawns, nonhumans and their places must be shielded with strict laws, real enforcement, and militant vigilance, if they are to survive.

6 To avert the impending mass extinction, the outward expansion and connection of nature's free spaces is imperative, by means of increasing the area of protected land and seas well beyond what is presently allotted—roughly 15 percent for land and 3 percent for the ocean. Vast portions of land and seas must be shielded from infrastructure, crop and animal agriculture, human settlements, mining projects, industrial fishing, and shipping lanes. The intertwined emergencies of a hemorrhaging biodiversity and catastrophically changing climate have inspired bold proposals from scientific and conservation communities.

7 One mandate is the cessation of all primary forest destruction—boreal, temperate, and tropical. The dry and wet tropical forests of the Americas, Africa, Asia, and Oceania harbor a stupendous diversity of known and unknown life-forms, and their conversion into poaching grounds, cattle ranches, soybean monocultures, oil palm plantations, and other

agricultural as well as mining ventures is a clarion sign of a global economy gone haywire and, by any standard of rationality, insane. There exists a general impression (encouraged, undoubtedly, by the view of forests as "renewable resources") that ancient forests can regenerate after exploitation. This is a convenient fiction, for the biodiversity of primary forests does not simply bounce back. In former agricultural land, for example, after forests regrow (if they are allowed to regrow), their native species diversity is diminished and their species composition is characterized by "an increase of common, competitive species at the expense of ancient forest indicator species". In other words, biotic homogenization prevails in post-agricultural ecologies, contradicting a widespread notion that forests recover from past agriculture or other ventures. As a global community, we must find the wisdom and the will to let ancient forests be: the most prestigious art museum of the world cannot hold a candle to them, unless its walls were to start breathing and its art objects were to come to life.

8 Another mandate is to vastly expand protected areas in the ocean, 97 percent of which is currently open to fishing and only 3 percent protected. Oceanographer Sylvia Earle calls for inverting this ratio—protecting 97 percent of the ocean. Indeed, a recently discussed proposal has been to stop all fishing in the high seas. Coupled with expanding protected areas along the coasts, full protection for the high seas would allow marine life to recover from the industrial plunder of the past century. Instituting this measure would signal our capacity to match our will to the knowledge of the marine devastation that continues under our watch. It would reflect our wisdom to treasure marine biodiversity for its own grandeur and for the sake of the experience and livelihood of future people.

9 Nature requires protection at such an enormous scale in order to redress the extinction and climate crises. The urgency of expanding and connecting protected areas today cannot be overstated. Continental-and oceanic-scale conservation—based on cores (such as parks, wilderness areas, marine protected areas, and restored places) and linked via corridors—constitutes the only effective means to staunch biodiversity losses, stop the nonhuman genocide of extinction, and soften the blows of extreme dryness, extreme wetness, extreme heat, and other extreme weather events that come with anthropogenic climate change. In the words of conservationists Kristine Tompkins and Tom Butler, the Half Earth vision, "puts the largest possible framing on contemporary conservation efforts. Parks and wilderness areas aren't just about scenery or recreation or campfire cookouts with the family during summer vacation. They are the fundamental building blocks of a durable future—for humanity and all of the other species who call this planet home". As E. O. Wilson writes, "there is no solution available ... to save Earth's biodiversity other than the preservation of natural environments in reserves large enough to maintain populations sustainably. Only Nature can serve as the planetary ark".

(Adapted from Crist, 2019)

New Words and Expressions

cauldron /ˈkɔːldrən/ *n.* 大锅
alchemy /ˈælkɪmɪ/ *n.* 炼金术
interpenetrate /ˌɪntəˈpenɪtreɪt/ *vt.* （互相）贯穿，渗透
uncharted /ʌnˈtʃɑːtɪd/ *adj.* 未知的
plenitude /ˈplenɪtjuːd/ *n.* 丰富
ramification /ˌræmɪfɪˈkeɪʃən/ *n.* 后果
discrete /dɪˈskriːt/ *adj.* 分离的
intersect /ˌɪntəˈsekt/ *v.* 交叉
teeter /ˈtiːtə/ *vi.* 摇摆
extinction /ɪkˈstɪŋkʃən/ *n.* 灭绝
rampage /ræmˈpeɪdʒ/ *n.* 狂暴行径
rank /ræŋk/ *adj.* 强烈的；极端的
arbitrary /ˈɑːbɪtrərɪ/ *adj.* 任意的
incite /ɪnˈsaɪt/ *vt.* 煽动
impunity /ɪmˈpjuːnɪtɪ/ *n.* （惩罚，损失，伤害等的）免除
mycorrhizal /ˌmaɪkəʊˈraɪzəl/ *adj.* 菌根的
pledge /pledʒ/ *vt.* 保证
rewild /ˌriːˈwaɪld/ *v.* 重新野化
biome /ˈbaɪəʊm/ *n.* 生物群落
indispensable /ˌɪndɪˈspensəbl/ *adj.* 不可缺少的
extermination /ɪkˌstɜːrmɪˈneɪʃən/ *n.* 消灭
militant /ˈmɪlɪtənt/ *adj.* 好斗的
vigilance /ˈvɪdʒɪləns/ *n.* 警觉
avert /əˈvɜːt/ *vt.* 防止
impending /ɪmˈpendɪŋ/ *adj.* 即将发生的
imperative /ɪmˈperətɪv/ *adj.* 重要的
hemorrhaging /ˈhemərɪdʒɪŋ/ *adj.* 严重流失的
mandate /ˈmændeɪt/ *n.* 命令
stupendous /stjuːˈpendəs/ *adj.* 巨大的
poaching /ˈpəʊtʃɪŋ/ *n.* 非法捕猎
monoculture /ˈmɒnəʊˌkʌltʃə/ *n.* 单作，单种栽培
clarion /ˈklærɪən/ *adj.* 清晰响亮的
homogenization /hɔˌmɔdʒənaɪˈzeɪʃən/ *n.* 同质化
redress /rɪˈdres/ *vt.* 纠正
staunch /stɔːntʃ/ *vt.* 止住，减少
genocide /ˈdʒenəʊˌsaɪd/ *n.* 大屠杀（尤指种族灭绝）

Exercises

Section I　Knowledge Focus

Task 1 ***Read the whole passage and complete the following diagram by filling in the blanks with words from the passage.*** （***ONE WORD ONLY***）

Concept of an ecological civilization:
➢ Earth as a 1) __________ planet
➢ a thriving 2) __________ of human and nonhuman communities

Overview of the problem:
➢ lack of 3) __________ for both nonhumans and humans
➢ the "4) __________ behavior", promoting greed, power, arrogance, and harm

Proposed solution:
➢ Restore 5) __________ to allow the biosphere to express its inherent nature.

Three pragmatic frameworks:

➢ Solution 1

- Protect 25%-75% of 6)_________ to preserve biodiversity.
- Increase and 7) _________ protected areas on land and sea, shielding them from human exploitation.

➢ Solution 2

- Stop 8)_________ forest destruction to prevent biotic 9)_________.

➢ Solution 3

- Expand ocean protection by 10)_________ the current protection ratio (97% protected, 3% open to fishing).
- Stop all 11)_________ sea fishing.

Urgency of action:

➢ Expand and connect conservation on continental and oceanic 12)_________ to redress extinction and 13)_________ climate crises.

➢ Only nature can serve as the planetary ark.

Task 2 ***Based on your understanding of the passage, decide whether the following statements are true or false. Put T for true and F for false in the blank provided before each statement.***

__________ 1) The passage equates wilderness to a "cauldron" in which Earth performs alchemical processes.

__________ 2) The rampage called "civilized behavior" is characterized in the passage as promoting values of humility and mutual respect.

__________ 3) The biosphere expressing its inherent nature is described as a "work of art that gives birth to itself".

__________ 4) The passage suggests that the notion of forests recovering from past agricultural endeavors is widely accurate and not contested.

__________ 5) One of the proposals for ocean conservation is to invert the current protection ratios, aiming for 97% protection.

__________ 6) The impending mass extinction can be averted solely by stopping primary forest destruction.

__________ 7) "Biotic homogenization" in the passage refers to an increase in common species at the expense of ancient forest indicator species after agricultural activities.

__________ 8) E. O. Wilson's perspective is that solutions other than the preservation of natural environments exist for saving Earth's biodiversity.

__________ 9) According to the passage, the urgency of conservation cannot be overstated, and it requires conservation on both continental and oceanic scales.

__________ 10) The article implies that the preservation of biodiversity and combating climate change should be seen as intertwined emergencies requiring bold proposals.

Section II Language Focus

Task 3 ***Match the term in the left column with an explanation given in the right column and write the corresponding letter in the space provided below.***

1) biosphere	A. *n.* fungal roots aiding plant communication
2) mycorrhizal networks	B. *n.* returning ecosystems to their natural state
3) rewilding	C. *n.* undisturbed, natural land
4) wilderness	D. *n.* Earth's living organisms and environments

1)__________ 2)__________ 3)__________ 4)__________

Task 4 ***Complete the following sentences with appropriate words or expressions given below. Change the form where necessary.***

abundant	conservation	dignity	ecological	immunity

1) No society, regardless of its technological advancements, has complete __________ to the consequences of environmental degradation, as history has shown.
2) Historically, our ancestors lived in harmony with the __________ natural resources, understanding the rhythms of the earth and its offerings.
3) Throughout the centuries, indigenous communities have championed __________ ethics, ensuring the sustainability of their lands for future generations.
4) Ancient civilizations, like the Mayans and Indus Valley communities, had an intrinsic understanding of __________ balance, designing their settlements in harmony with nature.
5) Traditional practices rooted in respect for the land not only emphasized environmental stewardship but also recognized the inherent __________ of all living beings.

impending	imperative	interpenetrate	intersect	restore

6) As the Roman Empire expanded, it became __________ to develop sophisticated aqueducts, showcasing their understanding of environmental management.
7) The Aztecs, understanding the importance of clean water, built intricate systems to __________ and purify their water sources.
8) The artworks of ancient China beautifully __________ the relationship between humans and their environment, revealing deep cultural reverence for nature.
9) Awareness of __________ deforestation in ancient cultures led to early rituals and traditions emphasizing the sacredness of trees and forests in their folklore.
10) In the heart of the Amazon Rainforest, where ancient tribal customs __________ with modern conservation efforts, researchers have discovered that centuries-old agricultural practices offer solutions to current environmental challenges.

Task 5 ***Find the words in the box that have the same meaning as the underlined words in the following sentences, and write the corresponding word in the space provided.***

discrete	extermination	indispensable	pledge	plenitude

__________ 1) Throughout environmental history, there have been distinct periods where societies recognized the importance of sustainable practices, each shaping their unique conservation strategies.

__________ 2) The wisdom passed down from indigenous cultures regarding land management has become essential in modern strategies for biodiversity conservation and ecosystem restoration.

__________ 3) At the historic environmental summit, nations from every corner of the globe came together to affirm their commitment to preserving ancient forests and their rich cultural histories.

__________ 4) The archaeological site revealed a(n) abundance of artifacts, shedding light on ancient civilizations' deep respect for nature and their sustainable ways of life.

__________ 5) The extinction of countless species due to rapid industrialization has prompted historians and ecologists to examine the cultural shifts that undervalued the symbiotic relationship between humans and nature.

Task 6 ***Match the words in the left column with those in the right column to form appropriate expressions. Then complete the following sentences with one of the expressions. Change the form or add articles where necessary.***

embrace	blow
incite	extinction
redress	killing
perform	alchemy
soften	truth

1) While it's easier to ignore the consequences of our actions, it's crucial to __________ about climate change to enact meaningful solutions.
2) Propaganda, when skillfully wielded by powerful entities, can __________ and promote hatred, as history has tragically shown.
3) Efforts to __________ of endangered species involve not just conservation but also understanding and rectifying the root causes of habitat loss.
4) Community support and intervention programs can __________ of economic downturns, ensuring that vulnerable families are not left destitute.
5) In the heart of the rainforest, nature is continuously __________, transforming sunlight and water into a flashing carpet of life, supporting an astonishing range of forms.

Section III　Sharing Your Ideas

This passage reflects on rewilding as an element of an ecological civilization—an act of urgency to protect and repair the biosphere of the Earth.

Task 7 ***As you watch the video, consider the following questions for the subsequent discussion.***

1) What is the impact of rewilding on local ecosystems and biodiversity?
2) The Half Earth vision proposes a grand scale of conservation. What are the key challenges in implementing this vision, and how can they be addressed?
3) What roles do governments, businesses, and local communities play in the realization of the Half Earth vision?
4) How can we balance the needs of human development and biodiversity conservation within the context of the Half Earth vision?

视听资源

After watching the video, we will discuss these questions and explore the ways in which we can contribute to the realization of the Half Earth vision, ensuring that we address both the imperatives of biodiversity conservation and the needs of human development.

Active Reading 2

Warming Up

视听资源

Task 1 ***Have you observed the changes occurring in our climate? Watch a short video clip and answer the following questions.***

1) What are the primary causes of climate change as discussed in the video?
2) Which effects are mentioned in the video?
3) Based on your knowledge, what steps can individuals take to reduce their carbon footprint?

视听资源

Task 2 ***Watch a short video and discuss the following questions with your partners.***

1) Is there any difference between industrial farming and agroforestry farming?
2) Which farming method do you think is more environmentally friendly?

Reading

Carbon Farming: Climate Change Solution or Greenwashing?

—European agriculture produces millions of tons of CO_2 every year. Could encouraging farmers to capture carbon on their land to sell credits to businesses help reduce emissions?

1 "Now society expects much more from farmers" reflects Belgian dairy farmer Kris Heirbaut. "Not only that we produce food, but that we also help reduce climate change".

2 Heirbaut owns a farm in the Flemish town of Temse, 30 kilometers (19 miles) from the port city of Antwerp. Outside the farm, he has a small store selling dairy products, including ice cream, made from milk from his own cows.

3 Two years ago, concerned with agriculture's damage to the environment, Heirbaut signed up to a "carbon farming" pilot project funded by the European Union that aims to improve agricultural soil health while tackling climate change.

4 The project, concluded in summer 2021, enabled farmers in Belgium, the Netherlands, Germany and Norway to sell carbon credits for carbon sequestered on their land. The EU gave the farmers scientific advice and administrative support to issue their first credits to local companies.

5 In December 2021, the EU presented its carbon farming initiative, with the intention to replicate the project across Europe. The EU initiative encourages farmers to make changes such as applying fertilizers rich in carbon, reducing tillage that disturbs the soil, and planting trees and crops that can absorb carbon dioxide from the atmosphere.

Changing Farm Practices

6 Soils are vital carbon stores, but industrial farming, rather than absorbing CO_2, often releases it into the atmosphere—for example through plowing which, if done repeatedly, can result in the degradation of the soil.

7 Since signing up for the initiative, Heirbaut has planted a field of narrow-leaf plantain—a perennial type of weed with high carbon sequestration potential—as well as crops that can rotate throughout the year. In total, he has about 14 hectares (34 acres) of land covered with grasses, clover, alfalfa, ribwort plantain and chicory, which can sequester CO_2 all year long.

8 "Because we mow four times a year, but do not need tillage machinery to work in the soil, all the carbon that the roots of the plants will bring into the soil will stay there", he explains.

9 Heirbaut also has a field dedicated to agroforestry, in which trees or shrubs are grown around crops and pasture. These trees sequester carbon, and the shade of the trees allows cows to graze on grass in the summer, another practice that can help farmers absorb CO_2.

Improving Soil Health

10 The EU hopes that giving farmers a financial incentive will help them increasingly shift more agricultural land from emitting carbon to capturing it. The carbon farming initiative is part of the European Green Deal, the EU's road map to become climate neutral by 2050. An estimated more than 385 million tons of CO_2 come from European farming, according to European Environment Agency data—just over 10% of the bloc's total emissions.

11 "Carbon content of soil is a good proxy for soil health", says Celia Nyssens of the European Environmental Bureau, a network of environmental NGOs. Europe's intensive farming practices have damaged soils over recent decades. A 2020 European Commission study found that around 60%-70% of EU soil is currently degraded, largely due to intensive farming, the use of pesticides or excessive irrigation.

12 No-till farming, such as that used by Heirbaut, is one way to improve soil health. Other techniques to help soils retain carbon include crop rotation, planting cover crops on fallow land to maintain the nitrogen in the soil and using compost instead of chemical fertilizers. These practices also protect other essential nutrients in the soil that plants need to grow, which in turn reduces the need for agrochemicals.

Carbon Scheme Criticisms

13 But carbon-offsetting schemes have long been criticized for allowing companies, individuals and states to simply buy their way to net-zero goals. In a letter to the US Congress last year, over 200 NGOs asked lawmakers to oppose a bill, currently under debate in the House of Representatives, that could see a carbon farming initiative set up in the US.

14 "Power plants, refineries and other polluters could purchase these carbon credits to offset their emissions, or even increase them, instead of actually reducing and eliminating them", the signatories argued.

15 Carbon farming has attracted several multinational corporations. Microsoft, for instance, has bought over $4 million (€3.6 million) in carbon credits generated from US farmers piloting carbon farming projects since 2021, to offset the tech giant's emissions.

16 But the companies using Heirbaut's greener farming practices to offset their pollution aren't multinationals. Earlier this year he sold his first carbon credits to Milcobel, a local dairy processor for roughly €50 per ton of CO_2 saved.

17 He hopes to collaborate with other small businesses in the Flanders region. "The advantages of buying carbon credits locally is that you can visit the farmers—people can sit and have a drink with us, visit the fields", he says. Although the pilot project has finished, Heirbaut intends to continue with carbon farming.

18 Farmers can physically sequester up to around 3.6 metric tons of carbon per hectare each year, according to a study commissioned by Rabobank (a Dutch bank). But to do this, they must make significant investments in changing their farming practices—as well as hiring independent experts to undertake expensive soil analyses to evaluate its health.

Getting Creative on the Farm

19 Heirbaut says it's a cumbersome process that could put some farmers off the scheme. Some critics fear it could make the benefits of carbon farming inaccessible to smaller operations, and favor larger industrial agriculture operations.

20 Carbon offsets generated from biofuel or reforestation projects have contributed to land grabbing—massive acquisitions of land usually from major corporations—across the world.

21 Nyssens of the European Environmental Bureau believes that a poorly designed EU carbon farming system risks falling into the same trap. "If we create a system where there is even more value from having land, because you can also sell credits from carbon sequestration, you will worsen those problems", she says.

22 But on his small dairy farm, Heirbaut says carbon farming gives him the opportunity to improve the health of his land, while making a little extra income. And it isn't his only eco-friendly venture. In addition to carbon farming, he's also building a lab to create new food products based on microalgae—protein-rich cells increasingly used as a substitute for meat.

23 "In the past decades, farmers have specialized in one thing, and now we know if this one thing goes wrong it can be a big problem", Heirbaut says, as he welcomes guests to his store and treats them to its latest release: an ice cream made of hazelnut and his homegrown microalgae.

(Adapted from Gumbau, 2022)

New Words and Expressions

greenwashing /ˈgriːnˌwɔʃɪŋ/ *n.* 漂绿(以环保为旗号的做法)

tillage /ˈtɪlɪdʒ/ *n.* 耕作

agroforestry /ˌægrəʊˈfɔrɪstrɪ/ *n.* 农林复合经营

agrochemical /ˌægrəʊˈkemɪkl/ *n.* 农用化学品;农药

proxy /ˈprɔksɪ/ *n.* 代理

offset /ˈɔfset/ *vt.* /*n.* 抵消; 补偿,抵消

signatory /ˈsɪgnətərɪ/ *n.* (协议的)签署者,签署方

cumbersome /ˈkʌmbəsəm/ *adj.* 笨重的

acquisition /ˌækwɪˈzɪʃən/ *n.* 获得,收购

sequestration /ˌsiːkwəˈstreɪʃn/ *n.* 隔离;封存

venture /ˈventʃə(r)/ *n.* 风险项目

microalgae /ˌmaɪkrəʊˈældʒiː/ *n.* 微藻

pilot /ˈpaɪlət/ *v.* 试点,(推广前)对……做试点;*n.* 试点项目

Exercises

Section I Knowledge Focus

Task 1 ***Read the whole passage and complete the following diagram by filling in the blanks with words from the passage.*** (***ONE WORD ONLY***)

Introduction	• CO_2 emissions : 1)_________ million tons estimated from European farming, accounting for over 2) _________ of the EU's total emissions. • Potential solution: Carbon farming.
EU Carbon Farming Projects	• EU carbon farming 3)_________ project: EU-backed support for farmers from 4 countries. • EU carbon farming initiative: Intends to replicate the pilot project across Europe, part of European Green Deal.
Changing Farm Practices	• Soils: Vital carbon stores. • Techniques adopted: Reducing tilling, or no-till farming, 4)_________ weeds, croprotation, agroforestry, planting cover crops on fallow land , and using 5)_________ instead of chemical fertilizers.
Improving Soiol Health	• Current soil degradation: caused by intensive farming, 6)_________ or excessive irrigation. • European Green Deal: Climate 7)_________ by 2050, and its intention to incentivize farmers financially. • Improved soil health: Carbon content of soil as a good 8)_________.
Criticisms of Carbon Scheme	• Concerns over 9) _________ goal shortcuts for polluters: argument against allowing polluters to 10) _________ emissions instead of reducing them. • Opposition from NGOs. • Example of Microsoft buying carbon credits.
Financial and Logistical Challenges	• Investment required for farmers: significant. • Soil analyses : Expensive. • Risk: Making carbon farming 11) _________ to smaller operations. • Land 12) _________ concerns: due to increasing land value through carbon credits.
Positive Impact on Farmers	• Example of Kris Heirbaut. • Selling carbon credits: 13) _________ /ton CO_2, up to 14) _________ metric tons/hectare annually. • Opportunity to improve land health. • Innovation in eco-friendly food products: e.g. 15) _________ -based food.

Task 2 ***Given below are five statements. Each statement contains information given in one of the paragraphs of the text. Identify the paragraph from which the information is derived. Answer the questions by writing down the paragraph number (1-23) for each statement.***

__________ A) A study suggests that farmers have the potential to capture and store a significant amount of carbon on their land each year.

__________ B) Conventional agriculture often releases carbon into the atmosphere instead of absorbing it like soils normally do, due to practices like regular plowing damaging soils over time.

__________ C) Heirbaut has adopted specific agricultural practices, including planting narrow-leaf plantain and implementing crop rotation, to enhance carbon sequestration on their land.

__________ D) The production of carbon offsets from biofuel or reforestation initiatives has resulted in the controversial practice of land grabbing, where corporations acquire vast amounts of land.

__________ E) The initiative's signatories claim that the ability to purchase carbon credits could enable polluting entities to compensate for or even increase their emissions rather than implementing actual emission reduction measures.

Section II Language Focus

Task 3 ***Match the term in the left column with an explanation given in the right column and write the corresponding letter in the space provided below.***

1) cover crops	A. *n.* decomposed waste providing nutrients for plants
2) compost	B. *n.* planting crops without soil disturbance
3) no-till farming	C. *n.* sequential planting for soil fertility and pest control
4) crop rotation	D. *n.* crops planted for soil protection and nutrient cycling

1) __________ 2) __________ 3) __________ 4) __________

Task 4 ***Identify what is being described in the following phrases. Choose you answer from the box below and write the corresponding word in the space provided.***

initiative	pilot	proxy	sequestration	signatory

__________ 1) capture and storage, specifically referring to carbon dioxide

__________ 2) substitute or representative acting on behalf of another

__________ 3) planned action or project to achieve a goal

__________ 4) a person or entity formally agreeing and supporting an agreement

__________ 5) experimental program to test feasibility or effectiveness

Task 5 ***Complete the following sentences with appropriate words or expressions given below. Change the form where necessary.***

acquisition	analyze	cumbersome	emit	greenwashing

1) The environmental scientists conducted detailed ___________ of water samples to assess pollution levels and identify contaminants.
2) The industrial plant ___________ toxic gases into the air, contributing to air pollution and climate change.
3) The environmental organization celebrated the ___________ of a large tract of land for conservation purposes.
4) The process of obtaining permits for renewable energy projects was often ___________, delaying their implementation.
5) The company's exaggerated claims of sustainability were exposed as ___________, hiding their actual environmental impact.

irrigation	offset	pilot	sequester	venture

6) The entrepreneur embarked on a sustainable farming ___________, combining aquaponics and organic practices.
7) The carbon emissions from the factory were ___________ underground, preventing their release into the atmosphere.
8) The municipality decided to ___________ an innovative green infrastructure project, aiming to evaluate its efficacy in urban stormwater management before considering a city-wide rollout.
9) The company purchased carbon ___________ to compensate for their greenhouse gas emissions.
10) Modern farmers often use drip ___________ a highly efficient method that delivers water directly to the roots of plants, minimizing waste and conserving water.

Task 6 ***Translate the following Chinese expressions into English with what you have learned from the text.***

1)低碳农业　2)气候变化减缓　3)碳封存　4)工业化养殖　5)土壤退化
6)碳抵消　7)土地侵占　8)减排　9)碳积分

Task 7 ***Translate the following paragraph from Chinese into English, using the phrases and expressions from Task 6.***

低碳农业是一种可持续的农业模式，它通过固碳和减少排放，在减缓气候变化方面发挥着至关重要的作用。与传统的工业化农业不同，低碳农业注重提升土壤健康状况，减轻土壤退化问题。借助碳补偿机制，农民可以通过低碳农业获得相应的碳信用。不过，在推行低碳农业的过程中需要应对土地掠夺等问题，以确保实施过程的公平性，切实保护当地社区的权益。总的来说，低碳农业在固碳、减排和促进农业可持续发展方面极具前景。

Section III Sharing Your Ideas

Carbon farming makes use of natural processes and sustainable agricultural practices to sequester carbon dioxide, offering a potential solution to combat climate change.

Task 8 ***Watch a short video to see how a farming family has put this into practice.***
In the news report, the scientist notes that the full effects of carbon farming are only observable over the long term and across extensive areas. So let's delve into the long-term benefits, challenges, and limitations.

Please consider the following questions in your sharing.

1) What are the potential long-term benefits of carbon farming?
2) What challenges might farmers encounter when implementing carbon farming practices?
3) How might the cost and investment required for carbon farming impact its long-term viability?
4) What can governments, organizations, and individuals do to help facilitate the adoption of carbon farming?
5) What are the potential environmental, economic, and social impacts of carbon farming in the long run?
6) How does carbon farming fit into broader strategies for combating climate change?

Part Three China's Environmental Story

Active Reading 3

Warming Up

视听资源

Task ***Watch a short video and discuss the following question with your partners.***
In what ways does the Kubuqi project demonstrate efforts to rejuvenate and enhance rural communities?

Reading

Benefits of China's Low-carbon Transition

1 The world currently faces the twin crises of COVID-19 and climate change. The COVID-19 pandemic has caused severe damage to the world economy, disrupting lives, societies and politics. Yet despite the magnitude of the crisis, which is truly global, the risks posed by unmanaged climate change are likely to be greater and longer-lasting than those posed by the pandemic. The threats must be dealt with simultaneously, but with recognition that we cannot go back to the old economic growth model and repeat the mistakes that followed the financial crisis of 2008-2010—of overinvesting in traditional high carbon infrastructure.

2 Sustainable growth is the only feasible path forward, and it is one of the few areas where a broad consensus might be reached globally. As China leads the world out of the COVID-19 crisis, it has a great opportunity to build a new vision for the country's own development and its relationship with the world. As one of the first major G20 countries to have made the transition from rescue to recovery, China can show the world that the crisis offers an opportunity for building back better, as recovery measures can also boost growth by accelerating the transition to the inevitable low-carbon economy.

3 In September 2020 the President announced at the United Nations General Assembly that China will aim for carbon neutrality by 2060. This significant pledge shows China's long-term ambitions and priorities, and that the Chinese government has linked low-carbon development and carbon-neutral transition with the country's sustainable development strategy and long-term prosperity. The benefits of the low-carbon transition manifest in many ways.

Low-carbon Transition Can Act as a New Driver of Growth

4 Sustainable growth, with the low-carbon transition at its core, is not only good for the environment and climate, but also can offer new drivers of growth and rural revitalisation, and lasting improvements to wellbeing. Sustainable growth does not mean expensive ways

of doing things; in fact, quite the opposite—it can push down the cost in two ways. On the one hand, it helps shift the production possibility boundary, or in economic terms the production possibility frontier (PPF), outwards (in terms of the movement of the PPF curve itself), through discovery and innovations. In other words, advancement in new, low-carbon technologies could increase a country's ability to produce more goods based on the same inputs. On the other hand, it can push the economy towards lower costs through the increasing returns to scale and through improved efficiency that achieves a certain level of output with fewer material inputs.

5 The new growth story, which will be innovation-driven and requires investment to achieve, will make China more competitive and much cleaner. Government commitment, by providing a clear sense of direction, can foster innovation and investment for sustainable growth and help manage the transition as efficiently and fairly as possible. The returns on investment will be strong, unlocking economic, social and environmental benefits.

Low-carbon Transition Can Facilitate Economic Upgrading

6 China's economic expansion in the past has relied heavily on input factors of labour, capital and resources including energy, land, water and minerals. With the continuous increase of energy consumption, problems arise, such as environmental pollution and energy resource insecurity, which have gradually become the main bottlenecks restricting further development of the nation's economy. The low-carbon transition is both a challenge and an opportunity, as it is to some extent consistent with the needs for new forms of growth, structural transformation and economic upgrading in China. Thus, well-managed low-carbon development strategies can contribute to climate change mitigation and serve as catalysts for upgrading China's economy.

7 Investment in sustainable growth and low-carbon innovations would help promote the structural transformation of industries towards higher skills and technologies. China is already a leader in technologies of the emerging green economy (e. g. batteries, solar PV cells, and electric vehicles). If structured well, this new growth pathway can accelerate technology innovations and lay the foundation for long-term economic prosperity and competitiveness.

Low-carbon Transition Offers Better Job Opportunities

8 China's energy transition will create millions of jobs in the renewable energy sector. A recent study suggests that the job-creation rate of renewable industries is between 1.5 and three times that of traditional energy industries. According to the International Renewable Energy Agency (IRENA) (2020), globally there were 11.5 million people employed in the renewable energy sector in 2019, with the solar photovoltaic (PV), bioenergy, hydropower and wind power industries being the biggest employers. Almost 40 per cent of global renewable energy jobs are in China, estimated at over 4 million in 2020. In contrast, the coal mining industry in China is in decline. The number of coal workers dramatically reduced from 4.5 million in 2015 to around 2.7 million in 2020.

Low-carbon Transition Supports the Economy Through Ensuring Energy Security

9 Many studies have suggested a strong positive correlation between economic growth and energy demand, highlighting the challenges associated with reducing energy demand to tackle climate change due to the potential negative impact on economic growth. But as energy demand continues to rise, while it is pressing to reduce consumption of fossil fuels and cut carbon emissions, energy security is becoming an increasingly important issue in many countries.

10 Failing to ensure energy security comes with significant economic consequences, including energy price shocks, physical availability of energy as a crucial input factor in production, and effects in the demand side of the economy.

11 Against this background, the low-carbon transition implies the need for a very great expansion of low carbon power from renewables, to reduce dependency on fossil fuels and thus improve national energy security, which is one of the top priorities for China. Renewables, including solar power and wind, are much safer than fossil fuels from the perspectives of energy security, human health and climate change. The role of renewables in improving energy security is increasingly recognised. Unlike most fossil fuels, in particular oil and natural gas, renewables do not have risks associated with political instability of energy producing countries. The promotion of renewable energy can improve the diversity of the overall electricity generation portfolio and thus support energy supply security. As an important part of China's energy resource endowment, renewables have become cost-efficient options of energy delivery.

(Adapted from Stern et al. ,2021)

New Words and Expressions

magnitude /ˈmæɡnɪtjuːd/ *n.* 重大;重要性

sustainable /səˈsteɪnəb(ə)l/ *adj.* 可持续的

consensus /kənˈsensəs/ *n.* 共识

infrastructure /ˈɪnfrəˌstrʌktʃə/ *n.* 基础设施

carbon neutrality /ˌkɑːbən njuːˈtræləti/ *n.* 碳中和

revitalisation /riːˌvaɪtəlaɪˈzeɪʃn/ *n.* 复兴

mitigation /ˌmɪtɪˈɡeɪʃn/ *n.* 减轻

catalyst /ˈkætəlɪst/ *n.* 促使变化的人;引发变化的因素

competitiveness /kəmˈpetətɪvnəs/ *n.* 竞争力

photovoltaic /ˌfəʊtəʊvɔlˈteɪɪk/ *adj.* 光伏的

hydropower /ˈhaɪdrəʊˌpaʊə(r)/ *n.* 水电

correlation /ˌkɔrəˈleɪʃən/ *n.* 相关性

endowment /ɪnˈdaʊmənt/ *n.* 天赋资源

Exercises

Section I Understanding the Text

Task 1 ***Discuss the following questions in small groups.***

1) Based on the passage, how might China's response to the 2008-2010 financial crisis have influenced its approach to the COVID-19 pandemic's economic aftermath?
2) How does the passage describe the impact of sustainable growth on the "production possibility frontier (PPF)"?
3) In the context of the passage, what is the essence of the relationship between low-carbon transition and economic prosperity in China?
4) Based on the passage, how does the decline of the coal industry in China correlate with the growth in the renewable energy sector?
5) Given the emphasis on energy security in the passage, why might the political stability of energy-producing countries be significant to China?

Section II Developing Critical Thinking

Sustainable development is a concept that refers to the kind of growth and progress that meets the needs of the present without compromising the ability of future generations to meet their own needs. Many people believe that it should be the cornerstone of modern growth, driven by the advancement of social productivity and scientific innovation. It is widely recognized as essential for addressing the global challenges of our time, offering a pathway to create solutions that benefit both people and the planet.

Task 2 ***In celebration of Earth Day, the environmental club at your university invites both Chinese and international students to step into the role of sustainability auditors. In this capacity, students will assess and evaluate the environmental impact and the efficacy of sustainable practices within designated spaces or communities on and off-campus. The club will recommend the best sustainability proposals to local organizations or the university's administration to turn them into actual projects.***

You may prepare your presentation by following the instructions below.

Step 1 Identify an area

- Choose a local area or community space individually or in groups.
- Jot down initial observations about the space, especially regarding its environmental impact or sustainability.

Step 2 Conduct the audit

- Examine the environmental and sustainability measures in your selected area.
- Capture images, if possible, to support your findings.

Step 3 Analysis & proposal

- Drawing on your notes and images, identify areas of improvement in the chosen space.
- You can refer to the passages in this unit:

Active Reading 1 emphasizes the importance of not taking resources for granted.

Active Reading 2 suggests agricultural or gardening practices that can be adopted, even in urban settings.

Active Reading 3 advocates for energy-saving measures or alternatives.

- Create a detailed proposal on:

  What changes can be implemented?

  How will these changes benefit the community and the environment?

  Potential challenges and solutions.

Step 4 Complete the checklist provided Below

Step 5 Class sharing

- Present your audits and proposals to the class.
- Please provide constructive feedback and engage in thoughtful discussion.
- Cast your vote for the most impactful sustainability proposals in the class.

Sustainability Audit Checklist

Audit Information

Name(s) ______________________

Date __________________________

Location /Area of Audit ______________

Observations

1. Resource Consumption

➢ Water

- Faucets

  Presence of aerators /sensors to minimize flow.

  Signs of leakage or wastage.

- Irrigation

  Efficiency of irrigation system (e. g. , drip vs. sprinkler).

  Presence of timers or moisture sensors to prevent over-watering.

- Water Storage

  Capacity and functionality of rain barrels /rainwater harvesting.

  Water recycling /reuse systems (e. g. , greywater systems).

- Usage Promotion

  Presence of educational signs promoting water conservation.

  Availability of water-efficient appliances (low-flow toilets, efficient washing machines).

2. Waste Management

➢ Trash

- Collection Bins

  Adequate number and placement.

  Labeling for correct waste disposal (e. g. , non-recyclable materials).

- Recycling

  Specific bins for paper, plastics, glass, metals.

  Cleanliness and proper segregation within bins.

- Composting

  State of compost (moisture, smell, color).

  Differentiation between green (nitrogen-rich) and brown (carbon-rich) composting materials.

3. Land Use and Agriculture

➢ Green Spaces

- Landscaping

  Percentage of native plants vs. introduced species.

  Wildlife-friendly features (birdhouses, pollinator gardens).

- Agriculture /Gardening

  Crop rotation methods used.

  Vertical gardening or space-saving methods.

4. Carbon Footprint and Emissions

➢ Transport

- Parking

  Designated areas for carpooling or eco-friendly vehicles.

  Quality and accessibility of bike storage facilities.

- Public Transport

  Proximity to bus stops or train stations.

  Information on public transport schedules.

5. Building and Infrastructure

➢ Material Sustainability

- Use of locally-sourced or recycled building materials.
- Presence of hazardous or non-sustainable materials.

➢ Energy Efficiency

- Windows double-glazed or treated for energy efficiency.
- Insulation type and effectiveness in walls, roofs, and floors.

6. Community Involvement and Education

➢ Engagement Initiatives

- Availability of workshops or clubs focused on sustainability.
- Presence of community gardens or shared sustainability projects.

➢ Information Accessibility

- Quality and quantity of informational boards about environmental practices.
- Regular updates or newsletters on sustainability advancements.

7. Additional Observations

1) ______________________________

2) ______________________________

3) ______________________________

8. Recommendations (Derived from readings and personal insights)

1) ______________________________

2) ______________________________

3) ______________________________

References

AUSTRALIAN GOVERNMENT DEPARTMENT OF SUSTAINABILITY, ENVIRONMENT, WATER, POPULATION AND COMMUNITIES, 2022. Feral european rabbit (*Oryctolagus cuniculus*) [R/OL]. [2022-06-30]. https://www. agriculture. gov. au/ sites/default/files/ documents/ rabbit. pdf.

BLACK B, LYBECKER D L, 2008. Great debates in American environmental history (Vol 1) [M]. Westport:Greenwood Press.

BULLARD R D, MOHAI P, SAHA R, et al., 2007. Toxic wastes and race at twenty: 1987—2007 [R/OL]. (2021-03-02) [2022-06-10]. https://www. ejnet. org/ej/twart. pdf.

CLAPP B W, 1994. An environmental history of britain since the industrial revolution[M]. London: Routledge.

CRIST E, 2019. Abundant earth: Toward an ecological civilization [M]. Chicago: University of Chicago Press.

ELVIN M, LIU T, 1998. Sediments of time: Environment and society in Chinese history [M]. Cambridge: Cambridge University Press.

GUMBAU A, 2022. Carbon farming: Climate fix or greenwashing? [EB/OL]. (2022-03-05)[2022-06-08] https://www. dw. com/en/carbon-farming-climate-change-solution-or-greenwash ing/a-61532175.

HASSAN F A, 2005. A river runs through Egypt: Nile floods and civilization[J/OL]. [2022-06-08]. http://www. agiweb. org/geotimes/apr05/feature_NileFloods. html.

HUGHES J D, 2009. An environmental history of the world: Humankind's changing role in the community of life [M]. London: Routledge.

HUGHES J D, 2015. What is environmental history? [M]. 2nd ed. Cambridge: Polity.

JI P Y, FAN Y, 2013. Academic english: An integrated course[M]. Beijing: Foreign Language Teaching and Research Press.

KIPLE K F, 2000. The cambridge world history of food (Volume 1) [M]. Cambridge: Cambridge University Press.

MANN C C, 2012. 1493: Uncovering the world created by Columbus [M]. Visalia: Vintage.

MARKS R B, 2011. China: Its environment and history [M]. Lanham: Rowman &

Littlefield Publishers.

MENG Y, DU J Y,2020. Irrigation system: Ancient flood management wisdom [EB/OL]. (2020-08-17) [2022-06-22]. https://news. cgtn. com/news/2020-08-17/Dujiangyan-Irrigation-System-Ancient-flood-management-wisdom-T1Y6L10Hok/index. html

Merchant C, 2005. The Columbia guide to American environmental history [M]. New York: Columbia University Press.

NATIONAL GEOGRAPHIC SOCIETY, 2022. How European rabbits took over Australia [EB/OL]. [2022-06-28]. https://education. nationalgeographic. org/resource/how-european-rabbits-took-over-australia/od-management-wisdom-T1Y6L10Hok/index. html.

PAN Y,2021. Developing socialist ecological civilization with Chinese characteristics [C/OL]// PAN J. Beautiful China: 70 Years Since 1949 and 70 People's views on eco-civilization construction. (2021-03-02) [2022-06-08]. https://link. springer. com/chapter/10. 1007/978-981-33-6742-5_39.

PONTING C, 2007. A new green history of the world: The environment and the collapse of great civilizations [M]. London: Penguin Books.

PRICE N, DU J Y, 2022. Irrigation system [EB/OL]. [2022-06-15]. https://environmentalchina. history. lmu. build/group-page-theme-2-water-control/dujiangyan-irrigation-system/.

ROLFE D, 1917. Environmental influences in the agriculture of ancient Egypt [J]. The American Journal of Semitic Languages and Literatures, 33(3):157-168.

ROSEN A, 2012. Effects of the fukushima nuclear meltdowns on environment and health [EB/OL]. (2012-03-09) [2022-06-25]. http://www. fukushimadisaster. de/fileadmin/user_upload/pdf/english/ippnw_health-effects_fukushima. pdf.

SMITHONIAN AMERICAN ART MUSEUM, 2015. The dust bowl [EB/OL]. (2015-02) [2022-06-18]. https://americanexperience. si. edu/wp-content/uploads/2015/02/ The-Dust-Bowl. pdf.

STACY J S, 2022. The American environmental movement: Surviving through diversity [EB/OL]. [2022-06-10]. https://webs. wofford. edu/whisnantdm/Sixties/Environment/Movement. pdf.

STEINBERG T, 2002, Down to earth: Nature's role in American history [M]. Oxford: Oxford University Press.

STERN N, XIE C P, 2022. China's new growth story: Linking the 14th Five-Year Plan with the 2060 carbon neutrality pledge [EB/OL]. (2022-05-10) [2022-06-10]. https://www. lse. ac. uk/ granth aminstitute /publication/ chinas-new-growth-story-linking-the-

14th-five-year-plan-with-the-2060-carbon-neutrality-pledge-2/.

SUSTAINABLE FOOTPRINT, 2022. Easter Island, a lesson for us all [EB/OL]. [2022-06-22]. https://sustainablefootprint. org/teachers/theme-lessons/easter-island-a-lesson-for-us-all/.

WANG W, 2020. Biodiversity conservation in China: A review of recent studies and practices [J/OL]. Environmental Science and Ecotechnology, (2020-05-10) [2022-06-10]. https://www. sciencedirect. com/science/article/pii/S266649842030017X.

WANG Z Q, 2024. China makes new successes in desertification prevention and control[N/OL]. People's Daily (2024-12-16) [2025-01-10] http://en. people. cn/mobile/new/content. html? cI=1003&nI=20254302&aT=m

ZHAO R N,2024. Recycled products shatter old ideas about ceramic waste, People's Daily Online, [N/OL], (2024-05-17) [2025-03-10] http://en. people. cn/n3/2021/0519/c90000-9851521. html.

Keys to Exercises

UNIT 1 An Introduction to Environment and History

Active Reading 1

Warming Up

Task

1) Meaning: If something is said to be as dead as a dodo, it is not working, obsolete, or unavailable. It refers to the flightless and now extinct bird that was taken by sailors to their ship and kept alive for fresh meat.
2) The Portuguese named the birds dodo, meaning stupid. Dodos were caught and eaten into extinction during the 17th century. Thereafter, the phrase passed into the English language.
3) The extinction of dodo bird is one of the famous examples of biodiversity loss due to massive human activities. This example shows how nature can be altered radically by human interference and how this human-nature interaction can place long-lasting effects on humans and future generations.

Exercises

Section I Knowledge Focus

Task 1

1) B 2) I 3) H 4) C 5) D
6) G 7) E 8) A 9) A 10) F
11) F

Task 2

1) E 2) D 3) C 4) A 5) B

Task 3

A) 7 B) 6 C) 3 D) 8 E) 1 F) 5 G) 4 H) 2

Section II Language Focus

Task 4

1) static 2) conserve 3) sustain 4) reserve 5) temperate 6) realm

Task 5

1) perceptions 2) realm 3) sustainable 4) temperate 5) conservative
6) ethic 7) tribe 8) devastate 9) artificial 10) constantly

Task 6

1) pristine wilderness 2) temperate zone 3) static force
4) aesthetic appeal 5) tribal rites

Task 7

Answers may vary.

Indians developed sophisticated cultures, social organizations and lingual systems in different environments throughout North America, surviving by hunting, gathering, fishing and horticulture long before the Europeans arrived in America. However, a direct result of European colonists' arrival was social and economic transformations. Indians were forced to leave their homelands and also suffered the loss of their cultures; the Indian population was decimated by disease and war. During these transformations, many Indian groups experienced great hardship and pain. Fortunately, Indian cultures and traditions have been preserved and still practiced in some communities.

Section III Sharing Your Ideas

Task 8

Answers may vary.

A few human-environment interactions examples are given below:

1. The extraction of oil and gas has numerous detrimental consequences for the environment. Oil spills have significant financial repercussions; they disrupt transportation and local; when birds are exposed to the oil spill, they lose their ability to seek food essential for their survival. Individuals are also exposed to unsafe seafood as a result of spills.
2. Due to expanding population and per capita consumption, water withdrawals have increased internationally over the previous three decades. Large-scale water extraction from bodies of water for domestic, agricultural, or industrial use limits the amount of water accessible to present and future generations. As a result, water demand rises, resulting in unsustainable consumption of water resource.

Active Reading 2

Warming Up

Task

1) Major environmental issues we face include, but are not limited to, climate change, deforestation, pollution, biodiversity loss, and resource depletion.
2) Answers may vary.
3) For some people, the book was a call to regulate substances capable of catastrophic harm. Others objected that Carson hadn't mentioned DDT's role in controlling the threat insects posed to human health.

Exercises

Section I Knowledge Focus

Task 1

1) F 2) T 3) F 4) T 5) T

Task 2

1) D 2) B 3) B 4) D 5) B

Task 3

1) Conservation 2) Environmentalism 3) founder 4) presidential
5) emergence 6) warned 7) pesticides 8) signed
9) subdiscipline 10) conservationists

Section II Language Focus

Task 4

1) logging 2) disappearance 3) wilderness 4) limitless
5) exploited 6) conservation 7) turned his back 8) dedicated
9) preserving 10) environmentalists

Task 5

1) merit 2) cease 3) apt 4) sound 5) customary

Task 6

1) disregard 2) campaigning 3) contaminated
4) denuded 5) deteriorating

Task 7

1）一场自觉的探索 2）学术尝试 3）惯常性哑谜 4）可再生资源
5）一次多层次的尝试 6）人为导致的气候变化 7）环保设施 8）前所未有的数字 9）残留性农药 10）全民意识

Section III Sharing Your Ideas

Task 8

1) Earth Day is an event which takes place every year on the 22nd of April. It was started to encourage people around the world to think about and learn what we can do to look after our planet.
2) Earth Day reminds us that we need to make changes to our lifestyles.
3) (Answers may vary) Picking up litter, planting trees and flowers, recycling, etc.

Active Reading 3

Warming Up

Task

Answers may vary.

Exercises

Section I Understanding the Text

Task 1

1) B 2) G 3) A 4) C 5) D 6) F 7) E

Section II Developing Critical Thinking

Task 2

Answers may vary.

UNIT 2 Human-Environment Interactions in Early Times

Active Reading 1

Warming Up

Task

1) That was the name given by the explorers because they believed they had landed in India.
2) They migrated from Asia, more than 15,000 years ago.
3) There are many reasons; many were killed by the invaders, and even more died of diseases brought to the new world by Europeans.
4) American Indians, also called Indians, Native Americans, Indigenous Americans, Aboriginal Americans, Amerindians, or Amerinds, are member of any of the aboriginal peoples of the Western Hemisphere.

The ancestors of contemporary American Indians were members of nomadic hunting and gathering cultures. These peoples traveled in small family-based bands that migrated from Asia to North America during the last ice age; from approximately 30,000-12,000 years ago, sea levels were so low that a "land bridge" connecting the two continents was exposed. Some bands followed the Pacific coast southward, and others followed a glacier-free corridor through the centre of what is now Canada. Although it is clear that both avenues were used, it is not certain which was more important in the peopling of the Americas. Most traces of this episode in human prehistory have been erased by millennia of geological processes: the Pacific has inundated or washed away most of the coastal migration route, and glacial meltwater has destroyed or deeply buried traces of the inland journey.

The earliest ancestors of Native Americans are known as Paleo-Indians. They shared certain cultural traits with their Asian contemporaries, such as the use of fire and domesticated dogs; they did not seem to have used other Old World technologies such as grazing animals, domesticated plants, and the wheel.

Archaeological evidence indicates that Paleo-Indians traveling in the interior of North America hunted Pleistocene fauna such as woolly mammoths (*Mammuthus* species), giant ground sloths (*Megatherium* species), and a very large species of bison (*Bison antiquus*); those traveling down the coast subsisted on fish, shellfish, and other maritime products. Plant foods undoubtedly contributed to the Paleo-Indian diet, although the periglacial environment would have narrowed their quantities and varieties to some extent. Plant remains deteriorate quickly in the archaeological record, which can make direct evidence of their use somewhat scarce. However, food remains at Paleo-Indian sites including Gault (Texas) and Jake Bluff (Oklahoma) indicate that these people used a wide variety of plants and animals.

Exercises

Section I Knowledge Focus

Task 1

1) stony 2) risky 3) warmer 4) conducive 5) excellent 6) favored

Task 2

1) fishing 2) gathering 3) venturing 4) migrating
5) farming 6) harvesting 7) sowing 8) packing

Task 3

1) T 2) F 3) F 4) F 5) F

Section II Language Focus

Task 4

1) C 2) D 3) A 4) B

Task 5

1) hardy 2) dormant 3) privation 4) exploit 5) underscored
6) proliferate 7) prune 8) deposited 9) tangled 10) attune

Task 6

1) bolster 2) compel 3) jeopordize 4) domesticated 5) conductive

Task 7

1) temperate climate 2) inviting target 3) cut down on pests 4) crop yields

Section III Sharing Your Ideas

Task 8

Answers may vary.

Active Reading 2

Warming Up

Task

1) The islant is the most remote inhabited place in the world and, there are giant stone statues all over it.
2) 887.
3) It provides many different explanations, including ecocide and the disruption and destruction caused by the European newcomers.
4) Answers may vary.

Exercises

Section I　Knowledge Focus

Task 1

1) volcanic　2) mammals　3) vegetation　4) rituals　5) monument
5) 7,000　7) deforested　8) agriculture　9) fishing　10) statues
11) cannibalism　12) meagre

Task 2

1) rats　2) erosion　3) declined　4) protein　5) caves　6) primitive　7) trapped
8) conflicts　9) warfare　10) Slavery　11) cannibalism

Section II　Language Focus

Task 3

1) B　2) C　3) A　4) E　5) D

Task 4

1) indigenous 2) topple　3) enchanting　4) regress　5) degradation
6) meagre　7) drastic　8) stranded　9) flimsy　10) squalid

Task 5

1)考古遗址　8) sweat potato
2)茂密的植被覆盖　9) crop yield
3)复杂的仪式　10) golden age
4)在巅峰时刻　11) draught animal
5)消耗资源　12) soil erosion
6)克服困难　13) household goods
7)以灾难告终　14) primitive society

Task 6

复活节岛的命运也可以成为现代世界的教训。像复活节岛一样，地球只有有限的资源来支持人类社会及其所有需求。像岛上的居民一样，地球上的人类没有可行的逃离地球的手段。世界环境如何塑造了人类历史，人们又如何塑造和改变了他们生活的世界？其他社会是否落入了与岛上的居民相同的陷阱？在过去的几千年里，人类成功地为越来越多的人口以及日益复杂与技术先进的社会获得了更多的食物和更多的资源。但是，相比岛上的居民，我们现在是否能成功地找到一种耗尽我们可用资源而不致命地的生活方式，或者我们的所作所为是否正在不可逆转地破坏我们的生命支持系统？

Section III　Sharing Your Ideas

Task 7

Answers may vary.

Task 8

Answers may vary.

Active Reading 3

Warming Up

Task

The woolly mammoth was known for its large size, fur, and imposing tusks. Thriving during the Pleistocene ice ages, woolly mammoths died out after much of their habitat was lost as Earth's climate warmed in the aftermath of the last ice age. The species is named for the appearance of its long thick coat of fur.

The saber-toothed tiger is one of the most well-known species of saber toothed cats from the genus *Smilodon*. This extinct cat was named for the pair of elongated teeth in its upper jaw. The saber tooth tiger roamed across North and South America during the Pleistocene Epoch. It went extinct approximately 10,000 years ago.

Exercises

Section II Understanding the Text

Task 1

1) That occurred as early as 9,500 to 8,800 years ago in the Yangzi River Valley, where some people began to cultivate rice. By eight thousand years ago in northern China, dry farming based on millet emerged.
2) Because China has a variety of geography, climate, and soils.
3) Five. These are the boreal forest [of the northeast], the temperate forest of northern China, the subtropical forest of central and southwestern China, and the tropical forest of southern China.
4) Because China has a vast number of ecosystems and ecological niches without any glaciers.
5) It began about 9,500 years ago, not just in China but in at least four other parts of the world as well.
6) The technologies of farming and settled agriculture.

Section II Developing Critical Thinking

Task 2

Answers may vary.

UNIT 3 Agrarian Civilization and Environment

Active Reading 1

Warming up

Task 1

1) The first writing system in the world is believed to be Sumerian cuneiform, developed in ancient Mesopotamia.

2) The earliest writing known to us dates back to around 3000 B. C. E. and was probably invented by the Sumerians, living in major cities with centralized economies in what is now southern Iraq. The earliest tablets with written inscriptions represent the work of administrators, perhaps of large temple institutions, recording the allocation of rations or the movement and storage of goods. Temple officials needed to keep records of the grain, sheep, and cattle entering or leaving their stores and farms, relying on memory became impossible. So, an alternative method was required and the very earliest texts consisted of pictorial representations of the items scribes needed to record (known as pictographs).

Task 2

1) They used to hunt wild animals and gather wild plants for food.
2) Flood forecasting and summer droughts were hugely daunting for them.
3) They invented methods to control the river, such as levees to block unexpected floods, gated ditches to divert floodwater for irrigation when needed, oxen to pull the plough for planting, and a calendar to forecast the floods.

Exercises

Section I Knowledge Focus

Task 1

1) puzzle 2) low 3) evaporated 4) salt 5) waterlogged 6) fallow
7) outweighed 8) sustainable 9) replacement 10) salinisation

Task 2

1) yields 2) cultivatable 3) surplus 4) vulnerable 5) impoverished

Task 3

1) T 2) T 3) F 4) F 5) T

Section II Language Focus

Task 4

1) C 2) A 3) D 4) B

Task 5

1) dwindling 2) desolated 3) exacerbate 4) retention 5) posing
6) tolerate 7) delta 8) deduced 9) flourished 10) painstaking

Task 6

1) storage 2) degradation 3) intervention 4) impoverished 5) cultivatable

Task 7

1) lay fallow 2) exacerbate deforestation 3) apply irrigation
4) undergoes salinization 5) poses dilemmas

Section III Sharing Your ideas

Task 8

Answers may vary.

Active Reading 2

Warming up

Task

1) Mud-made houses.
2) Fruit, vegetable and wheat for baking.
3) They would still rely on hunting animals and gathering plants to survive, constantly facing food shortages, and they would not have other means of sustenance such as farming for self-sufficiency, because the land they lived on would not be suitable for farming without the Nile River.

Exercises

Section I Knowledge Focus

Task 1

1) Valley 2) agriculture 3) nourishment 4) unpredictable 5) water
6) management 7) bucket 8) measurement 9) sustainability 10) revolutions

Task 2

1) inartificial 改为 artificial
2) pharaohs 改为 laborers
3) extreme climate 改为 breaking dynasties
4) overuse 改为 disuse
5) beyond 改为 less than

Task 3

A) 9 B) 2 C) 6 D) 8 E) 3

Section II Language Focus

Task 4

1) took advantage of 2) floodwater 3) expected 4) ploughed 5) preserved

Task 5

1) famine 2) insulation 3) unrest 4) innovation 5) outgrowth

Task 6

1) tenacious 2) oriented 3) sacred 4) devised 5) exotically
6) ceased 7) sponsor 8) silted 9) dredged 10) instalments

Task 7

1) drainage system 2) flood-prone region 3) organic/ecological/biological farming
4) ecological damage 5) crop diversity 6) sustainable agriculture
7) water scarcity/crisis/stress 8) soil erosion
9) combat desertification 10) soil degradation

Section III Sharing Your ideas

Task 8

Answers may vary.

Active Reading 3

Warming up

Task

1) Because they were abundant, durable, and sustainable. Bamboo provided flexibility, while cobblestones added stability, thus effectively controlling water flow and preventing erosion with minimal resources.
2) Answers may vary.

Exercises

Section I Undestanding the Text

Task 1

1) At Dujiangyan, there is an ancient irrigation system constructed about 2000 years ago, which helps flood control.
2) Every design of the Dujiangyan irrigation system takes full advantage of the local environmental characteristics.
3) The main features are: the simplicity of the design, construction, and tools as well as the use of the natural and traditional materials.
4) It is the consistency of harmony between man and nature.

Section II Developing Critical Thinking

Task 2

Answers may vary.

UNIT 4 Changes in Biosphere in Early Modern Period

Active Reading 1

Warming Up

Task 1

1) Christopher Columbus.
2) The explorer Christopher Columbus made four trips across the Atlantic Ocean from Spain: in 1492, 1493, 1498 and 1502. He was determined to find a direct water route west from Europe to Asia, but he never did. Instead, he stumbled upon the Americas. Though he did not "discover" the so-called New World—millions of people already lived there—his journeys marked the beginning of centuries of exploration and colonization of North and South America.

Task 2

1) The trade of spices and silk with India and China was incredibly important and profitable, but the journey, whether over land or by sea, was long and dangerous. If he could succeed in sailing directly from Europe to Asia by going west, he could make a lot of

money and become very rich.

2) On August 3rd, 1492.

3) He brought back word of these new lands to Europe and began a new period of trade and colonization; plants, animals, people and diseases went back and forth across the ocean and had a lasting impact on almost every culture on the planet.

Exercises

Section I Knowledge Focus

Task 1

1) silk 2) Vikings 3) village 4) 1492 5) occupation

Task 2

1) choking 2) original 3) regrowth 4) extinction 5) release
6) viruses 7) immunity 8) innumerable 9) removed 10) demographic

Task 3

1) T 2) F 3) F 4) F 5) T

Section II Language Focus

Task 4

1) D 2) C 3) A 4) B

Task 5

1) calamity 2) swarmed 3) turbulent 4) consequential 5) legacy
6) depredation 7) initiate 8) engulf 9) ricocheted 10) culprit

Task 6

1) exceed 2) sugary 3) innumerable 4) scorch 5) expedition

Task 7

1) demographic catastrophe 2) scale insect
3) epidemic disease 4) ecological release

Section III Sharing Your Ideas

Task 8

Answers may vary.

Active Reading 2

Warming Up

Task

1) The Australian government lost the battle, because rabbits are good hole-diggers and can easily burrow under the fence and escape through tunnels.

2) Besides their lack of natural predators on the continent, rabbits are extremely adaptive; all they need is soil that is fit to burrow and short grasses to graze on, which are fairly easy to come by. Moreover, their success is aided by quick breeding: they can reproduce at a young age, and all year round, and give birth to more than four litters a year with five

kits in each.

Exercises

Section I　Knowledge Focus

Task 1

1) foxes　2) exotic　3) dingos　4) aggressive　5) same
6) extinction　7) decline　8) livestock　9) regeneration　10) grazing
11) topsoil　12) bedrock

Task 2

1) adaptive　2) burrow　3) habitats　4) offspring　5) breed
6) produce　7) unfavorable

Task 3

A) 9　B) 7　C) 4　D) 2　E) 6

Section II　Language Focus

Task 4

1) C　2) A　3) E　4) B　5) D

Task 5

1) erratic　2) exponentially　3) exotic　4) prolifically　5) arid
6) roam　7) accommodate　8) forage　9) extermination　10) prey

Task 6

1) unlikely areas　2) declared war　3) lay eggs　4) unfavorable conditions
5) produce litter　6) adaptable creatures

Task 7

在过去的150年里，澳大利亚尝试了许多方法来阻止兔子数量的增长。立法机构甚至通过了一些法律，其中包括要求土地所有者在他们的土地上诱捕、毒死或杀死兔子。由于兔子数量庞大，每一次试图抑制兔子数量增长的尝试都注定以失败告终。第二次世界大战后，澳大利亚对兔子宣战，并释放了一种致命的病毒，杀死兔子的效率为99.8%。最终，澳大利亚人认为他们找到了“灵丹妙药”。然而，存活下来的0.2%的兔子对病毒免疫。随着这些免疫兔子的繁衍，它们将免疫能力传递给它们的后代，新的免疫兔子的数量激增到数百万只。澳大利亚人似乎注定要与数百万只兔子共存很长一段时间。

Section III　Sharing Your Ideas

Task 8

Answers may vary.

Active Reading 3

Warming Up

Task

Answers may vary.

Exercises

Section I Understanding the Text

Task 1

1) The origin can be traced to the wild soybean plant *Glycine soja* that grew in abundance in northeastern China.
2) Soybeans were said to cause flatulence and were viewed mostly as a food for the poor during years of bad harvests.
3) A famine occurred, meaning not only too many mouths to feed but also the rise of the price of millet.
4) Because the wild soybean was sensitive as to the amount of daylight it required, and because the length of growing seasons varied from region to region.
5) Because as vegetarians, they were always interested in new plants foods and drinks.
6) Not only Buddhist priests but also soldiers, merchants, and travelers.
7) Because botanists were thrilled to study and classify soybean for experimental purposes.
8) Because during the American Civil War, shipping was disrupted, soybeans were frequently substituted for coffee, and soybeans were also cultivated as a forage plant. What's more, the new Department of Agriculture (USDA) played an invaluable role in promoting agricultural research, regulating the industry, and serving as an information generator for farmers.

Section II Developing Critical Thinking

Task 2

Answers may vary.

UNIT 5 Industrial Revolution and Environment

Active Reading 1

Warming Up

Task 1

1) The "London Smog Incident", which occurred in London from December 5 to 9, 1952, was known as one of the top ten environmental pollution disasters of the 20th century.
2) From December 7 to 13, 1873, a heavy fog shrouded London, causing nearly a thousand deaths, and marking the first recorded large-scale death related to smoke in history. The most severe tragedy occurred in 1952. From December 5 to 9, high pressure settled over London, preventing the exhaust gases from a large amount of factory production and residents' coal-fired heating from dissipating, causing them to accumulate over the city. The concentration of pollutants in the air continued to rise, many people experienced chest tightness, suffocation and other discomforts, leading to a sharp increase in both the incidence and mortality rates. During the five-day period of heavy fog, according to official UK statistics, more than 5,000 people lost their lives, and over 8,000 people

died within the two months after the fog cleared. This event, known as the "London Smog Incident", has become one of the top ten environmental pollution crises of the 20th century.

Task 2

1) In December 1952 a fog settled over the city for 5 days. When it dissipated, more than 150 people had been hospitalized, and more than 4000 people had died.

2) Coal burning.

Exercises

Section I Knowledge Focus

Task 1

1) coal 2) industrial 3) alkali 4) brine 5) trade 6) carbonic 7) ironwork
8) hydrochloric acid 9) fibres 10) crops

Task 2

1)F 2)T 3)F 4)T 5)T

Section II Language Focus

Task 3

1) condense 2) miscellaneous 3) fibre 4) emission 5) fabric

Task 4

1) pollutants 2) notorious 3) agents 4) trickled 5) dissolve
6) discharging 7) noxious 8) mainstay 9) solution 10) evaporate

Task 5

1) arouse interest 2) dilute national identity 3) corrosive impact
4) distinctly different 5) stand alone

Task 6

1) emitted 2) impurity 3) combination 4) readily 5) unwanted

Task 7

1) summed up 2) aroused little concern 3) saw fit to 4) contributed to

Section III Sharing You Ideas

Task 8

Answers may vary.

Active Reading 2

Warming Up

Task 1

George Perkins Marsh was born on March 15, 1801, in Woodstock, Vermont. Growing up in a rural setting, he retained a love of nature throughout his life. He occupied an important place in the history of the conservation movement and was known to many as the father of the environmental movement. In his masterpiece, *Man and Nature*, Marsh discussed the importance of a harmonious relationship between humans and the natural world, and warned of the dangers that could come if we used up natural resources. Marsh cautioned against exploiting natural resources when most people considered them to be infinite. He argued that man was doing great damage to the environment, which was ahead of its time. Most people of the era simply could not, or would not, grasp the concept that mankind could harm the Earth.

Task 2

1) Some historians believe that Britain's industrialization resulted from its position in global trade networks and its leadership of a vast empire.

2) Factors such as money, food, resources and markets may have contributed to the early industrialization in Britain.

3) The Industrial Revolution helped Britain become a global power and establish the largest territorial empire in history.

Exercises

Section I Knowledge Focus

Task 1

1) Dramatic 2) Adverse 3) upsetting 4) mismanaging
5) disturbing 6) discords 7) taming 8) weighing
9) accordingly 10) consumption 11) inception 12) soil conservation
13) unspoiled

Task 2

1) The natural environment was adversely impacted in immediately visible ways by the dramatic increase in industry. For example, the machinery of many factories was fueled by coal, causing smokestacks to belch black smoke into the air, and industrial byproducts flowed into the waterways, leaving them polluted.

2) The idealism of much of New England transcendentalism was interested in conservation or primitivism; while Marsh's view of transcendentalism advocated taming wilderness and for practical informed decisions and increased command over nature.

3) They became alarmed and began to search for ways to create a balance between industrial progress and the preservation of natural resources.

4) Because Marsh believed that nature, if not disturbed by humans, was only impacted by geological influences which may be regarded as constant and immutable. In that case, nature may retain its unchanging permanence of form, outline, and proportion.

5) Marsh argued that the natural environment was adversely impacted in immediately visible ways by a dramatic increase in industry, which was upsetting the natural balance of nature. Man was everywhere a disturbing agent because his activities destroyed the harmonies of nature. Marsh acknowledged the need for human use of the natural environment, but he thought humans should weigh the results and then act accordingly, rather than misuse or mismanage the natural resources.

Section II　Language Focus

Task 3

1) orientation　2) onset　3) accordingly　4) reverence　5) contradict

Task 4

1) dramatic　2) fuel　3) advocates　4) upset　5) ultimately

6) proportions　7) shattered　8) ominous　9) thrive　10) obsession

Task 5

1) aimed　2) taking away from　3) control　4) unchangeable　5) disturb

Task 6

1) vocal criticism　2) industrial byproduct　3) natural resources

4) reach/draw/come to/get to a conclusion　5) natural phenomenon

6) interdependence of environmental and social relationships

Task 7

1) 然而，随着人们越来越认为大自然的美丽值得歌颂和敬仰，一些观察者越来越不愿意忽视麻木不仁者对它的破坏。

2) 19 世纪，随着城市的发展，美国工业急剧增长。而随着工业的发展，自然环境直接受到了明显的不利影响。

3) 新英格兰超验主义的理想主义对保护自然或原始主义感兴趣，而马什的超验主义观点却主张驯服荒野。

4) 马什认为，工业的发展正在破坏自然平衡。这种影响的规模和范围让佛蒙特州政治家马什等见多识广的观察家们不知所措。

5) 在承认人类利用自然环境的必要性的同时，马什在他的《人与自然》一书中指责美国人滥用自然资源，对大自然的馈赠管理不善。

Section III　Sharing Your Ideas

Task 8

Answers may vary.

Active Reading 3

Warming Up

Task

Most discarded ceramics end up in landfills because they don't decompose. Some are repurposed for construction, like roadbeds or drainage systems, but large-scale recycling remains limited. To reduce waste, ceramics could be crushed and reused in new materials, or industries could design more sustainable products.

Exercises

Section I　Unstanding the Text

Task 1

1) Ceramic waste does not degrade. Landfilling takes up space, and glazed ceramics may

release harmful chemicals, polluting air and water.

2) Recycled bricks absorb and store water, reducing urban heat and improvingstorm water management, making them ideal for sponge cities and green infrastructure.

3) Market mechanisms encourage companies to invest in recycling, while regulations set pollution standards and enforce compliance. Both are needed for large-scale sustainability.

4) Reducing ceramic waste requires greater awareness, responsible consumption, and waste-conscious design in manufacturing and dailyuse.

Section II Developing Critical Thinking

Task 2

Answers may vary.

UNIT 6 Environmental Problems

Active Reading 1

Warming Up

Task 1

1) The Dust Bowl is a large area in the southwestern part of the Great Plains of America devastated by a series of dust storms due to a combination of drought and soil erosion during the 1930s and 1940s. The nationwide drought that began in 1930 shifted its center to the Great Plains after 1931, turning the area into a region plagued by dust storms.

2) The Great Plains in the Midwest of the United States were originally typical grassland ecosystems, which was referred to as the "Great American Desert" by early colonizers. But the federal government encouraged farmers to explore this "desert" during the westward expansion that began in 1862. The American farmers removed local weeds which could prevent wind and sand, and then planted wheat-based crops. This period of reckless reclamation is also known as the Great Plow-up. From the 1930s, pioneers paid a heavy price for their recklessness. An extended drought, high winds, and loss of millions of tons of topsoil resulted in the worst environmental disaster in American history. A large area was devastated by a series of dust storms, spreading from the Great Plains all the way to Chicago, New York City, and Boston. The air turned to earth and the storm carried tons of dust across America. The dust was so severe, so thick, over hundreds of miles that people for days didn't see the light. The storms forced about 250,000 people to leave the Great Plains, being one of the largest-scale migrations in American history.

Task 2

1) Dust storms hit these cities in 1934.

2) The dust storms were born out of a 100 million-acre dead zone, 2000 miles away in Oklahoma, Texas, Kansas and Colorado.

3) The environmental cataclysm involved an extended drought, high winds, and the loss of

millions and millions of tons of topsoil.

Exercises

Section I Knowledge Focus

Task 1

1) dust storms 2) Great Plains 3) drought 4) soil erosion 5) plowed
6) wheat production 7) machinery 8) visibility 9) halted 10) miserable
11) respiratory 12) contour lines 13) minimize 14) relief
15) record-breaking

Task 2

A) 5 B) 3 C) 1 D) 4 E) 3

Section II Language Focus

Task 3

1) manual 2) sweep 3) erosion 4) crack 5) terrace

Task 4

1) shrivelled/shriveled 2) fine 3) respiratory 4) raging 5) ravage
6) penetrated 7) combat 8) sustenance 9) yield 10) coincide

Task 5

1) disaster 2) resist 3) serious 4) initial 5) proportional

Task 6

1) migration 2) consequently 3) revolutionized 4) visibility 5) stimulus

Task 7

1) crippling effect 2) ecological balance 3) halt global warming
4) seal wounds 5) triple the odds

Section III Sharing Your Ideas

Task 8

Answers may vary.

Active Reading 2

Warming Up

Task 1

The Fukushima Daiichi nuclear disaster was an energy accident at the Fukushima Nuclear Power Plant, primarily initiated by the tsunami that was triggered by the Tohoku earthquake (a magnitude 9.0 earthquake in Northern Japan) on March 11, 2011. It is the largest nuclear disaster since the Chernobyl disaster of 1986 and was categorized as a Level 7 nuclear accident on the International Nuclear Event Scale (INES).

The disaster caused enormous contamination to the environment. The four large explosions, the fire in the spent fuel pond, smoke, evaporation of seawater used for cooling and deliberate venting of the pressurized reactors all led to the emission of radioactive isotopes into the atmosphere. The soil and marine environment were also contaminated. The radioactive substances, such as iodine-131 and caesium-137, were detected even in food and drinking water, the levels of which exceeded the radioactivity limits set for food and drink.

Task 2

1) Scientists recorded the magnitude of the earthquake at 8.9.

2) It was centered under the sea about 130 kilometers east of Sendai.

3) The earthquake created a tsunami wave ten meters high, which washed away boats, cars and houses in coastal areas north of Tokyo. It also led to tsunami warnings across the Pacific. The tsunami devastated entire neighborhoods in Sendai.

4) The quake happened about 370 kilometers northeast of Tokyo. It shook buildings in the capital, halting all train and subway traffic, and leaving many people unable to get home.

Exercises

Section I Knowledge Focus

Task 1

1) hit 2) Eastern 3) severely 4) pressure 5) radioactive 6) evacuation
7) cool off 8) contaminated 9) level 7 10) highest

Task 2

1) fire 2) fallout 3) bone-marrow 4) malignant 5) spent fuel
6) evaporated 7) single 8) pollutants 9) genetic mutation 10) exceeded

Task 3

1) T 2) F 3) T 4) T 5) F

Section II Language Focus

Task 4

1) tsunami 2) massive 3) secondary 4) ultimately 5) dose

Task 5

1) vent 2) contributor 3) incorporate 4) regardless 5) controversial
6) discharge 7) comprehensive 8) evacuated 9) pumping 10) deliberate

Task 6

1) constituted a threat 2) initial shyness 3) relieve anxiety
4) substantial impacts 5) desperate need

Task 7

在食品和饮用水中，不存在任何安全剂量的放射性物质。即使是最微量的放射性物质，也有可能导致基因突变和癌症。福岛核电站的熔毁致使日本食品和饮用水遭受严重污染。据国际原子能机构(IAEA)称，地震发生一周后，在茨城县和福岛县采集的几乎所有蔬菜和牛奶样本中，碘-131 和铯-137 的含量都超过了日本针对食品和饮用水设定的放射性限值。灾难发生后的数月里，在某些食品中发现的污染程度甚至更高。

Section III Sharing Your Ideas

Task 8

Answers may vary.

Active Reading 3

Warming Up

Task 1

There are some ways to prevent the loss of biodiversity and promote its increase. Firstly, endangered animals can be housed in wildlife parks, which can protect them. Secondly, captive breeding programs can be used to increase their populations. These programs can help the general public to form an emotional connection to these endangered species. Thirdly, we can protect endangered animals in their natural habitats by creating protected areas like national parks.

It's not just animals, plants are also at risk, so it's vital for us to protect them too. Besides creating the protected areas, we can protect the endangered plants by collecting seeds and storing them in seed banks. If a plant does become extinct, the seeds can be grown to restore its population.

Task 2

1) Because the greater biodiversity, the more secure all life on earth is, including ourselves.
2) They can lock away carbon and keep the climate stable.
3) Because in the last 50 years, human activities have destroyed habitats, reduced populations of wild animals by 60%, and even driven entire species extinct. For example, the number of lions in Africa has dropped by 65%, that of flying insects in Europe by 75%, that of Bluefin and tuna in the Pacific by 95%.

Exercises

Section I Understanding the Text

Task 1

1) Biodiversity refers to the ecological complex formed by living organisms (including animals, plants, and micro-organisms), their surrounding environment, and the sum of related ecological processes, including ecosystem diversity, species diversity, and genetic diversity.
2) Chinese scientists have made significant contributions to the conservation and sustainable use of biodiversity. For example, China increased the population of giant pandas by establishing nature reserves, and succeeded in artificially breeding them. Yuan Longping developed the first strain of hybrid rice in 1970, making outstanding contributions to the world's food security. Tu Youyou, a Nobel Prize winner, discovered *Qinghaosu* (*artemisinin*) that helped save millions of lives globally, especially in developing countries.
3) We need to further promote the mainstreaming of biodiversity conservation in decision-making and management at all levels of the government, and promote the integration of

conservation effectiveness into the assessment system of government officers.

4) Chinese ecological restoration programs have generated positive ecological effects by increasing vegetation cover, enhancing carbon sequestration, and controlling soil erosion.

5) China initiated several large-scale wetland restoration projects to convert reclaimed low-yield croplands back to wetlands.

Section II Developing Critical Thinking

Task 2

Answers may vary.

UNIT 7 Growth of Environmentalism

Active Reading 1

Warming Up

Task 1

1) Answers may vary. For example, I used a citronella-based mosquito repellent for my backyard. It helped reduce mosquito presence during outdoor gatherings.

2) In summary, pesticides are chemicals intended to kill, repel, or affect the behavior of any pest but very often pose impacts on the environment and human health. These chemicals can be sources of pollution to both the aquatic and soil ecosystems. Most pesticides are not that biodegradable and can stay long enough in the environment to bioaccumulate in animals up the food chain. Direct exposure to pesticides in human beings poses a number of health problems that include acute effects like skin irritations and nausea, plus chronic effects like cancers, endocrine disruption, and neurological disorders. Besides, pesticides make up food, drinking water, and air in an indirect way, such that it contributes to the pollution of the food we consume daily, water that one takes regularly, and finally, the air, as one requires.

Task 2

1) *Silent Spring*. It is about the misuse of chemicals and their toll on nature and human health, and about how human actions threaten the balance of nature.

2) Its author is Rachel Carson.

3) Her book drew impassioned dissent and vicious personal attacks on her.

4) Her work drew attention to how human actions, such as the use of pesticides, were harming the environment, and galvanized a generation of environmental activists, paving the way for the environmental movement to gain momentum in the US, leading to significant events and policies, such as the establishment of the Environmental Protection Agency (EPA) in the 1970s.

Exercises

Section I Knowledge Focus

Task 1

1) industrialization 2) efficient 3) Division 4) preservationists 5) independent 6) pollution 7) social 8) pollutants 9) infusion 10) mobilization 11) chemical 12) mainstream 13) Grassroots 14) anti-environmental 15) citizen

Task 2

1) F 2) F 3) F 4) T 5) F

Section II Language Focus

Task 3

1) B 2) D 3) A 4) C

Task 4

1) Decentralization 2) delineates 3) concomitant 4) exploitation 5) conservative 6) regulatory 7) litigate 8) lobby 9) supervision 10) mobilization

Task 5

1) ecompass 2) construe 3) imperil 4) salient 5) mobilization

Task 6

1) implemented policies 2) delineated plans 3) sounded alarm bells
4) marked the beginning 5) impact the consciousness

Section III Sharing Your Ideas

Task 7

Answers may vary.

Active Reading 2

Warming Up

Task 1

Three single-use plastics popped up in my day today: a plastic water bottle, a shopping bag, and packaging material. The sustainable items that could replace these would be a water bottle that can be refilled, cloth or jute bags for shopping, and articles with little packaging or biodegradable packaging.

Task 2

1) The speech discusses plastic's environmental impact and its disproportionate effects on marginalized communities.
2) The speaker cites plastic production pollution, health risks in low-income groups, and toxic disposal in developing countries.
3) "Cancer Alley" is a Gulf area with high petrochemical industry concentration, named for elevated local cancer rates.
4) The speaker links disposability to social injustices faced by poorer communities due to plastic.
5) Biomimicry is designing solutions inspired by nature's processes and systems.

Exercises

Section I Knowledge Focus

Task 1

1) pesticides	2) interconnected	3) urgent	4) hazards
5) underprivileged	6) banned	7) Third	8) victims
9) agroecosystems	10) small	11) grassroots	
12) protected	13) maturity		

Task 2

1) T 2) F 3) F 4) F 5) F

Section II Language Focus

Task 3

1) biodiversity 2) agroecosystem 3) sustainability 4) agroecology 5) globalization

Task 4

1) intrinsically 2) migration 3) acceded 4) Justice 5) globalization 6) myriad
7) underprivileged 8) sediment 9) unwitting 10) under-consumption

Task 5

1) take on a particular urgency
2) victims of environmental injustice
3) send a tide
4) application of agroecological principles
5) enacted by small farmers
6) fragmented landscapes
7) intrinsically interconnected
8) grassroots movements
9) biodiversity conservation
10) maturity of the Environmental Justice Movement

Task 6

In the face of escalating environmental challenges, the need for biodiversity conservation takes on a particular urgency. It is not just about preserving nature but also about addressing the victims of environmental injustice, who bear the brunt of these challenges. The application of agroecological principles, enacted by small farmers, plays a pivotal role in creating sustainable and resilient agricultural practices. These efforts are part of broader grassroots movements that strive for a more equitable and environmentally sound world. As these movements grow, they send a tide of change across fragmented landscapes, highlighting how environmental protection and social equity are intrinsically interconnected. Through these actions, the maturity of the Environmental Justice Movement becomes evident, marking a significant step towards recognizing and solving the complex web of issues facing our planet.

Section III Sharing Your Ideas

Task 7

Answers may vary.

Active Reading 3

Warming Up

Task

By planting trees one at a time and leveraging advanced technologies, Chinese have pushed back vast stretches of deserts and turning them into thriving forests. This transformation not only showcases their resilience and team effort but also represents a significant achievement in environmental restoration. Their dedication and determination are truly commendable.

Exercises

Section I Understanding the Text

Task 1

1) Countries need to focus on strong government policies, long-term investment, community involvement, and adapting afforestation strategies to local environmental and geographic conditions.
2) This shift helps improve decision-making by incorporating scientific research and systematic governance, allowing for more effective and sustainable environmental policies.
3) Afforestation combats desertification, enhances biodiversity, and increases carbon sequestration. It can also create jobs, improves local livelihoods, and boosts forest-based industries.
4) China's leadership in desertification control strengthens its soft power and fosters international collaboration on sustainability.
5) Answers may vary.

Section II Developing Critical Thinking

Task 2

Answers may vary.

UNIT 8 Towards a Sustainable Future

Active Reading 1

Warming Up

Task 1

1) Answers may vary. The lush forest probably evokes such emotions as tranquility, awe, and optimism, while the deforested area may make people feel desolated, alarmed, or guilty.
2) Healthy forests serve as habitats to countless species and, therefore, are crucial in

ensuring biodiversity. They absorb carbon dioxide and help in the conservation of both water and soil. Thus, to the environment, deforestation means the loss of habitats, a reduction in biodiversity, and the release of the stored carbon back into the atmosphere.

Task 2

1) Mass extinction is defined as a period during which there is a significant and widespread decline in the variety and abundance of life on Earth. The loss of species occurs rapidly and on a large scale over a geologically short period.
2) These extinctions were caused by various factors, including volcanic activity, climate change, asteroid impacts, sea levels changes, oceanic acidification, and other environmental disruptions.
3) The key element that might lead to the sixth mass extinction is believed to be human activity.

Exercises

Section I Knowledge Focus

Task 1

1) living 2) coexistence 3) dignity 4) civilized 5) wilderness
6) biomes 7) connect 8) primary 9) homogenization 10) inverting
11) high 12) scales 13) anthropogenic

Task 2

1) T 2) F 3) T 4) F 5) T 6) F 7) F 8) F 9) T
10) True.

Section II Language Focus

Task 3

1) D 2) A 3) B 4) C

Task 4

1) immunity 2) abundant 3) conservation 4) ecological 5) dignity
6) imperative 7) restore 8) interpenetrate 9) impending 10) intersect

Task 5

1) discrete 2) indispensable 3) pledge 4) plenitude 5) extermination

Task 6

1) embrace the truth 2) incite killing
3) redress the extinction 4) soften the blows
5) performing alchemies

Section III Sharing Your Ideas

Task 7

Answers may vary.

Active Reading 2

Warming Up

Task 1

1) The cause of current climate change is largely human activity, such as the increased gas concentration of greenhouse in the Earth's atmosphere and the burning of fossil fuels.
2) The video discussed the effects in 4 aspects, including ocean, weather, food and health. As temperatures rise, glaciers and ice caps melt, causing sea levels to rise. Climate change disrupts weather patterns, leading to more frequent and severe extreme weather events. Growing crops becomes more difficult. The area where animals and plants can live shift, and water supplies are diminished. In urban areas, the warmer weather creates an environment that traps and increases the amount of smog, causing problems such as asthma, heart problems, and lung cancer.
3) Answers may vary. We can use energy-efficient appliances, walk or bike, eat a plant-based diet, conserve water, recycle, reduce waste, buy sustainably, and support eco-friendly policies to lower our carbon footprint.

Task 2

1) Industrial farming, also known as conventional or intensive farming, is characterized by large-scale monoculture, heavy machinery usage, and the application of synthetic fertilizers and pesticides. Agroforestry farming is one type of carbon farming. It involves integrating trees or shrubs with crops and/or livestock on the same land.
2) Agroforestry farming is more environmentally friendly. Industrial farming typically involves extensive soil disturbance through plowing, which can contribute to soil erosion and carbon loss. In contrast, agroforestry provides biodiversity, soil conservation, carbon sequestration, water management, and sustainable agricultural production, thus mitigating climate change and protecting natural resources.

Exercises

Section I Knowledge Focus

Task 1

1) 385 2) 10% 3) pilot 4) perennial 5) compost 6) pesticides 7) neutrality
8) proxy 9) net-zero 10) offset 11) inaccessible 12) grabbing
13) ■50 14) 3.6 15) microalgae

Task 2

A) 18 B) 6 C) 7 D) 20 E) 14

Section II Language Focus

Task 3

1) D 2) A 3) B 4) C

Task 4

1) sequestration 2) proxy 3) initiative 4) signatory 5) pilot

Task 5

1) analyses 2) emits 3) acquisition 4) cumbersome 5) greenwashing

6) venture 7) sequestered 8) pilot 9) offsets 10) irrigation

Task 6

1) carbon farming
2) climate change mitigation
3) sequester carbon
4) industrial farming
5) soil degradation
6) carbon offsets
7) land grabbing
8) emissions reduction
9) carbon credits

Task 7

Carbon farming, a sustainable agricultural practice, plays a crucial role in climate change mitigation by sequestering carbon and reducing emissions. Unlike conventional industrial farming, carbon farming focuses on enhancing soil health and mitigating soil degradation. Through carbon offsets, farmers can earn carbon credits by implementing carbon farming practices. However, it is essential to address challenges such as land grabbing to ensure equitable implementation and protect local communities. Overall, carbon farming offers a promising approach for sequestering carbon, reducing emissions, and promoting sustainable agriculture.

Section III Sharing Your Ideas

Task 8

Answers may vary.

Active Reading 3

Warming Up

Task

The project creates jobs, raises incomes, involves locals as shareholders, and transforms the desert into productive land, boosting the rural economy and environment.

Exercises

Task 1

1) China recognized the need for a different economic growth model after the financial crisis and saw the pandemic as an opportunity to implement this change.
2) Sustainable growth shifts the PPF outward, indicating the potential for more production possibilities through discoveries and innovations.
3) Economic prosperity in China depends upon a successful low-carbon transition.
4) As the renewable energy industry expanded, it surpassed the coal sector, leading to its decline.
5) Political instability can lead to supply chain disruptions, affecting energy availability and prices.

Task 2

Answers may vary.